100

Advances in Biochemical Engineering / Biotechnology

Series Editor: T. Scheper

Advances in Biochemical Engineering/Biotechnology

Series Editor: R. Scheper

Recently Published and Forthcoming Volumes

Gene Therapy and Gene Delivery Systems
Volume Editors: Schaffer, D., Zhou, W.
Vol. 99, 2005

Sterile Filtration
Volume Editor: Jornitz, M.W.
Vol. 98, 2005

Marine Biotechnology II
Volume Editors: Le Gal, Y., Ulber, R.
Vol. 97, 2005

Marine Biotechnology I
Volume Editors: Le Gal, Y., Ulber, R.
Vol. 96, 2005

Microscopy Techniques
Volume Editor: Rietdorf, J.
Vol. 95, 2005

Regenerative Medicine II
Clinical and Preclinical Applications
Volume Editor: Yannas, I.V.
Vol. 94, 2005

Regenerative Medicine I
Theories, Models and Methods
Volume Editor: Yannas, I.V.
Vol. 93, 2005

Technology Transfer in Biotechnology
Volume Editor: Kragl, U.
Vol. 92, 2005

Recent Progress of Biochemical and Biomedical Engineering in Japan II
Volume Editor: Kobayashi, T.
Vol. 91, 2004

Recent Progress of Biochemical and Biomedical Engineering in Japan I
Volume Editor: Kobayashi, T.
Vol. 90, 2004

Physiological Stress Responses in Bioprocesses
Volume Editor: Enfors, S.-O.
Vol. 89, 2004

Molecular Biotechnology of Fungal β-Lactam Antibiotics and Related Peptide Synthetases
Volume Editor: Brakhage, A.
Vol. 88, 2004

Biomanufacturing
Volume Editor: Zhong, J.-J.
Vol. 87, 2004

New Trends and Developments in Biochemical Engineering
Vol. 86, 2004

Biotechnology in India II
Volume Editors: Ghose, T. K., Ghosh, P.
Vol. 85, 2003

Biotechnology in India I
Volume Editors: Ghose, T. K., Ghosh, P.
Vol. 84, 2003

Proteomics of Microorganisms
Volume Editors: Hecker, M., Müllner, S.
Vol. 83, 2003

Biomethanation II
Volume Editor: Ahring, B. K.
Vol. 82, 2003

Biomethanation I
Volume Editor: Ahring, B. K.
Vol. 81, 2003

Process Integration in Biochemical Engineering
Volume Editors: von Stockar, U., van der Wielen, L.A.M.
Vol. 80, 2003

Microbial Production of l-Amino Acids
Volume Editors: Faurie, R., Thommel J.
Vol. 79, 2003

Biotechnology for the Future

Volume Editor: Jens Nielsen

With contributions by
S. Arnold · M. A. Asadollahi · U. T. Bornscheuer · J. Maury
K. Møller · J. Nielsen · R. M. Raab · M. Reuss · J. Schmid
M. Siemann-Herzberg · G. Stephanopoulos · K. Tyo · C.-J. Yue
J.-J. Zhong

Springer

Advances in Biochemical Engineering/Biotechnology reviews actual trends in modern biotechnology. Its aim is to cover all aspects of this interdisciplinary technology where knowledge, methods and expertise are required for chemistry, biochemistry, micro-biology, genetics, chemical engineering and computer science. Special volumes are dedicated to selected topics which focus on new biotechnological products and new processes for their synthesis and purification. They give the state-of-the-art of a topic in a comprehensive way thus being a valuable source for the next 3–5 years. It also discusses new discoveries and applications. Special volumes are edited by well known guest editors who invite reputed authors for the review articles in their volumes.
In references *Advances in Biochemical Engineering/Biotechnology* is abbreviated as *Adv Biochem Engin/Biotechnol* as a journal.
Visit the *Adv Biochem Engin/Biotechnol* home page at springeronline.com

Library of Congress Control Card Number 2005925868

ISSN 0724-6145
ISBN-10 3-540-25906-6 **Springer Berlin Heidelberg New York**
ISBN-13 978-3-540-25906-0 **Springer Berlin Heidelberg New York**
DOI 10.1007/b136409

Springer is a part of Springer Science+Business Media
springeronline.com

Printed in Germany

Cover design: KünkelLopka GmbH, Heidelberg; *Design & Production* GmbH, Heidelberg
Typesetting and Production: LE-TEX Jelonek, Schmidt & Vöckler GbR, Leipzig

Printed on acid-free paper 02/3141xv – 5 4 3 2 1 0

Advances in Biochemical Engineering/Biotechnology Also Available Electronically

For all customers who have a standing order to Advances in Biochemical Engineering/Biotechnology, we offer the electronic version via SpringerLink free of charge. Please contact your librarian who can receive a password for free access to the full articles by registering at:

springerlink.com

If you do not have a subscription, you can still view the tables of contents of the volumes and the abstract of each article by going to the SpringerLink Homepage, clicking on "Browse by Online Libraries", then "Chemical Sciences", and finally choose Advances in Biochemical Engineering/Biotechnology.

You will find information about the

- Editorial Board
- Aims and Scope
- Instructions for Authors
- Sample Contribution

at springeronline.com using the search function.

Attention all Users of the "Springer Handbook of Enzymes"

Information on this handbook can be found on the internet at springeronline.com

A complete list of all enzyme entries either as an alphabetical Name Index or as the EC-Number Index is available at the above mentioned URL. You can download and print them free of charge.

A complete list of all synonyms (more than 25,000 entries) used for the enyzmes is available in print form (ISBN 3-540-41830-X).

Foreword

The "Volume 100" anniversary gives me the chance to reflect on our successful series "Advances in Biochemical Engineering/Biotechnology".

With the support of Springer, Professor Fiechter and Professor Ghose decided in 1971 to "stimulate many more talents to contribute through basic research dissemination of knowledge to the yet to emerge hybrid discipline of biochemical engineering". The thirty-five fruitful, dynamic, and remarkable years of this series are proof of the strength of their vision.

Advances in Biochemical Engineering/Biotechnology has provided a tremendous impetus on research and fruitful scientific exchange in this area. While the number of journals, series and books on biotechnology grew exponentially in this period, the "Advances" were always on the forefront of research and knowledge. Looking back, it is surprising that many of the "hot topics" of today were initiated in the early seventies (e.g. plant cell biotechnology, Vol. 2 (1972) and mammalian cell technology, Vol. 3 (1974)).

The initial idea to review the actual state of knowledge in this series was and is still a key for success for the Advances. While the first volumes contained different chapters with actual topics in biotechnology, we decided in 2000 to organize special topic volumes. Thus, each volume is dedicated to a certain area of modern biotechnology. I am proud that we were even able to convince the founding editors to contribute to these special volumes: Prof. A. Fiechter: History of Biotechnology (Vol. 69, 70 in 2000) and Prof. T.K. Ghose: Biotechnology in India (Vol. 84, 85 in 2003). I heartily thank both in the name of our scientific community for having this wonderful dream about this Springer series.

I also would like to take this opportunity to thank the editorial board members for their excellent work, and to express my appreciation to all of the authors. Over 50 well known scientists served on the board to improve the quality of this series from one volume to the next, and more than 2000 authors are responsible for the profound success of the Advances in Biochemical Engineering/Biotechnology series. It was a pleasure to cooperate with them and learn from their work and contributions.

My special thanks to the Springer team, and especially to Dr. Marion Hertel and Ulrike Kreusel. No volume could have been published without their endurance, helpfulness and friendliness. I am looking forward to the bright future of biotechnology and many more years of "Advances in Biochemical Engineering/Biotechnology".

Hannover, July 2005 Thomas Scheper

Preface

Even though fermentation processes have been carried out since the beginning of the 20th century, biochemical engineering was first established as a field during the development of the industrial penicillin fermentation process. Before this process was implemented the fermentation industry was based on relatively primitive bioreactors and many new developments were necessary in connection with establishing a large-scale process for the production of the first pharmaceutical derived from a fermentation process. In particular, the requirements for large volumes of sterile air and for large power inputs led to completely new bioreactor designs. Later there were many developments in design of an optimal fermentation strategy, and biochemical engineers contributed significantly to the large increase in productivity and yield obtained over the years, e.g., through introducing advanced fed-batch fermentation. All these developments led to the establishment of biochemical engineering as a separate discipline. At this time, a biochemical engineer was traditionally a chemical engineer that received some training in the life sciences.

In the 1970s there was a rapid development in life sciences; many new enzyme processes were developed, many biochemical engineers started to construct detailed mathematical models for the biocatalyst, and biological waste water treatment was introduced extensively. At this time Springer started the book series *Advances in Biochemical Engineering* (*/Biotechnology* was added later), and in the first volume in 1971, Fiechter and Ghose (the founding editors) stated, "Biochemical engineering is concerned with the effective practical application of the knowledge amassed by the biological sciences." Along these lines, the first volume contained chapters on (1) the nature of fermentation fluids; (2) cell separation; (3) enzyme kinetics; (4) production and application of enzymes; (5) overproduction of microbial metabolites and enzymes; (6) production of biomass from hydrogen and CO_2; and (7) production of single-cell protein from hydrocarbons. The editors expressed a hope that the book series would assist in the development of the "yet to emerge hybrid discipline of biochemical engineering."

Since then we have realized a biological revolution. Over the years there has been a shift in focus among biochemical engineers from development of new unit operations towards the use of engineering principles to understand the molecular mechanisms governing a given process and to use this knowledge for improving the process. This has particularly been manifested in the design of new enzymes through protein engineering and in the design of new cell factories for the production of a wide range of compounds. Thus, biochemical engineering today encompasses a wide range of activities and it would not be possible to include all aspects of modern biochemical engineering in one volume. However, the essence of biochemical engineering is basically the same as defined in 1971, namely *quantitative analysis of biological processes and use of this information for design of new molecules, cells, unit operations or complete processes.*

One area that has attracted much attention is the use of cell factories for the production of a wide range of chemicals. This area illustrates very well the definition of biochemical engineering specified above, and it has therefore been chosen as the focus of this volume. The use of cell factories for production of chemicals dates back to the production of penicillin, ethanol, acetone–butanol, glycerol and further to enzymes and other antibiotics. All these products were traditionally produced by cell factories that naturally produce these compounds. With the introduction of genetic engineering it has, however, become possible to combine traits from different organisms into a single cell factory, and thereby obtain a better production process – an approach referred to as metabolic engineering. This approach has been further supported by developments in genomics, which has enabled gaining much more detailed insight into the function of living cells at the systems level. This volume therefore contains chapters on concepts of metabolic engineering (by Stephanopoulos and coworkers), on detailed modeling of cellular systems (by Reuss and coworkers), on the use of plant cells for production of natural products (by Zhong and coworker), on the production of chemicals through the use of enzyme catalysis (by Bornscheuer), on the production of chemicals through the use of microbial cell factories (by Nielsen and coworkers), and finally on the control of fermentation processes through advanced monitoring techniques (by Ulber and coworkers). I hope you will enjoy reading the chapters and actively contribute to the further advancement of biochemical engineering in the future.

Lyngby, May 2005 Jens Nielsen

Contents

Adv Biochem Engin/Biotechnol (2005) 100: 1–17
DOI 10.1007/b136411

Published online: 5 July 2005

Metabolic Engineering

R. Michael Raab · Keith Tyo · Gregory Stephanopoulos (✉)

Department of Chemical Engineering, Room 56-459,
Massachusetts Institute of Technology, Cambridge, MA 02139, USA
gregstep@mit.edu

Abstract Metabolic engineering is a powerful methodology aimed at intelligently designing new biological pathways, systems, and ultimately phenotypes through the use of recombinant DNA technology. Built largely on the theoretical and computational analysis of chemical systems, the field has evolved to incorporate a growing number of genome scale experimental tools. This combination of rigorous analysis and quantitative molecular biology methods has endowed metabolic engineering with an effective synergism that crosses traditional disciplinary bounds. As such, there are a growing number of applications for the effective employment of metabolic engineering, ranging from the initial industrial fermentation applications to more recent medical diagnosis applications. In this review we highlight many of the contributions metabolic engineering has provided through its history, as well as give an overview of new tools and applications that promise to have a large impact on the field's future.

Keywords Metabolic engineering · Bioinformatics · Systems biology

1 Introduction

Metabolic engineering emerged with the advent of recombinant DNA technology [1]. For the first time it was possible to recombine genes from one organism with those of another, opening the door to a realm of possibilities not yet explored. While the initial applications of genetic engineering were simply producing human proteins in bacteria for therapeutic treatment of specific protein deficiencies, engineers quickly realized the vast potential of using multiple genes to create entirely new pathways that could produce a wide range of compounds from a diverse substrate portfolio [2, 3]. Aided by advanced methods for the analysis of biochemical systems, metabolic engineers set out to create new industrial innovations based on recombinant DNA technology.

Metabolic engineering is different from other cellular engineering strategies because its systematic approach focuses on understanding the larger metabolic network in the cell. In contrast, genetic engineering approaches often only consider narrow phenotypic improvements resulting from the manipulation of genes directly involved in creating the product of interest. The need for a systematic approach to cellular engineering has been demonstrated by several vivid examples in which choices for improving product formation, such as increasing the activity of the product-forming enzyme, have only resulted in incremental improvements in output [4, 5]. Intuitively, this makes sense. A typical cell has evolved to catalyze thousands of reactions that serve a multitude of purposes critical for maintaining cellular physiology and fitness within its environment. Thus changing pathways that do not improve fitness, or even detract from fitness within a population, often causes the cell's regulatory network to divert resources back to processes that optimize cellular fitness. This may lead to relatively small improvements in product formation despite large increases in specific enzymatic activities. Without a good understanding of the metabolic network, further progress is often difficult to achieve and must rely on other time-consuming methodologies based on rounds of screening for the phenotype of interest. Classical strain improvement (CSI) relies on random mutagenesis to accumulate genomic alterations that improve the phenotype. This method typically has diminishing returns for a variety of reasons: 1) it does not extract information about the location or nature of the mutagenesis; 2) it often results in deleterious mutations and therefore is less efficient, and; 3) it does not harness the power of nature's biodiversity by mixing specialized genes between organisms. Gene shuffling approaches attempt to correct the second and third issues by swapping large pieces of DNA between different parental strains to eliminate deleterious mutations or incorporate genes from other organisms. In contrast, metabolic engineering approaches embrace techniques that fill the gaps left by CSI and gene-shuffling methodologies by placing an emphasis on understanding the mechanistic features that genetic modifications confer, thereby adding knowledge that can be used for rational approaches while searching the metabolic landscape.

Metabolic engineering overcomes the shortcomings of alternative approaches by considering both the regulatory and intracellular reaction networks in detail. Research on the metabolic pathways has primarily focused on the effect of substrate uptake, byproduct formation, and other genetic manipulations that affect the distribution of intracellular chemical reactions (flux). Because many of the desired products are organic molecules, metabolic engineers often concentrate their efforts on carbon flow through the metabolic network. In diagnosing the metabolic network, engineers rely on intracellular flux measurements conducted in vivo using isotopic tracers as opposed to simply using macroscopic variables such as growth rate and metabolite exchange rates. The latter measurements contain less information about the intracellular reaction network and therefore give a very limited perception of the phenotype of the cell. Enzymatic assays can also provide helpful, but potentially misleading, information about the activity of an enzyme in the cell and cannot be used to calculate individual fluxes, which also depend upon the size of the metabolite pools and other intracellular environmental factors. Research on regulatory networks has ranged widely from engineering allosteric regulation, to constructing new genetic regulatory elements such as promoters, activators and repressors that influence the reaction network [6–8]. By understanding the systemic features of the network, metabolic engineering can identify rational gene targets that may not be intuitive when relying upon extracellular or activity measurements alone.

In practice metabolic engineering studies proceed through a cycle of perturbation, measurement, and analysis (Fig. 1). Measurement requires the ability to assay large parts of the network to extract as much information about the effect of an imposed network perturbation as possible. Gas

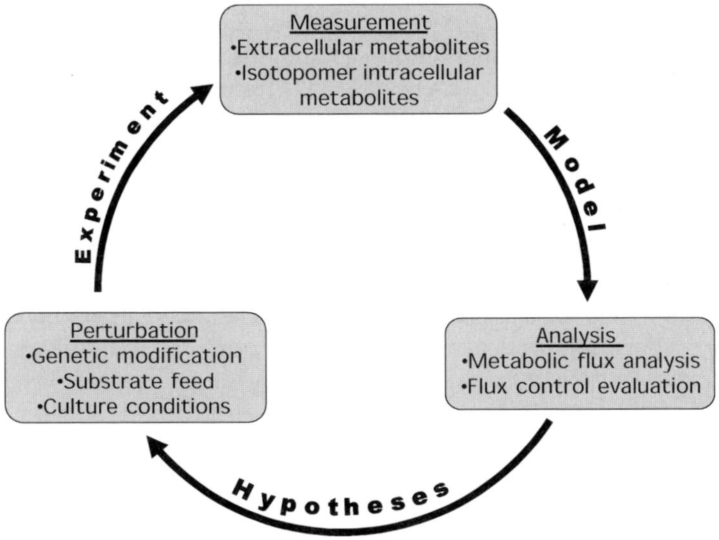

Fig. 1 The iterative approach of metabolic engineering. Metabolic engineering is an information-driven approach to phenotype improvement that involves (1) measurement, (2) analysis, and (3) perturbation. Data from measurements can be used to formulate models. These models can then be analyzed to generate new targets for manipulation (hypotheses). After performing the genetic manipulations, experiments must be formulated to determine how the metabolic network has adjusted to each genetic manipulation. The cycle can then continue, providing more information with each round

chromatography-mass spectrometry (GC-MS), and nuclear magnetic resonance (NMR) are commonly used to measure metabolite pools and the rates of chemical reactions within cells. Microarrays have been developed, and new proteomic tools are evolving, to monitor the response of gene expression to different perturbations. Finally, to complete the cycle before proceeding to the next iteration, robust analyses are necessary to determine which portions of the network are the most sensitive or amenable to genetic manipulation and to the generation of meaningful hypotheses from the vast quantities of data that can be gathered. By analyzing the differences in the metabolic fluxes following a perturbation, new targets can be identified that are most likely to improve the phenotype. The new targets set the foundation for hypotheses, leading to another perturbation of the network. Such perturbations are followed by another round of measurement and analysis and may include: increasing the activity of desirable enzymes within a pathway either by overexpression or deregulation, deleting enzymes that divert carbon to undesired byproducts, using different substrates, or changing the overall state of the cell to favor certain pathways.

As in other engineering sciences, metabolic engineering requires rigorous measurements to quantify cellular physiology. The *metabolic phenotype*, or movement of carbon through the reaction network of the cell, is a comprehensive measure of the cell's physiological state. The metabolic phenotype can be assessed using a variety of strategies. Extracellular metabolite uptake and production-rate measurements provide limited information about the intracellular reaction rates or fluxes. The existence of parallel pathways and branch points in the intracellular reaction network prohibits the determination of all fluxes using only extracellular measurements. Thus, to estimate the intracellular fluxes, these extracellular measurements must be complemented by knowledge of the intracellular reaction network and isotopic tracer measurements. By using stable isotopes (^{13}C) to label various positions within a substrate molecule, one can track the movement of carbon within the metabolic network. Performing these experiments in vivo generates the information necessary to obtain a more complete picture of the cellular response to a perturbation, allowing the engineering of the network as desired.

2
Applications

Metabolic engineering principles have had an impact on numerous areas within biology; however, its most common employment has been in developing new microorganism strains with tailored traits for bioprocessing and biocatalysis. The systematic treatment of an organism with multiple inputs, outputs, and chemical reactions defining its behavior enables metabolic engineers to optimize new traits efficiently for industrial applications. Many of the characteristics endowed to these new strains address some common bioprocessing challenges: 1) nonexistent or low product titer or yield, 2) expensive production substrate, and 3) excess byproduct synthesis. If these challenges can be met using metabolic engineering, the economics of the processes can often be substantially improved, leading to the financially competitive commercialization of new products from recombinant DNA technology.

Among the industrially relevant products of fermentation and cell culture that have been targets for metabolic engineering are citric acid [9], synthetic drug intermediates [10], ethanol [11], lactic acid [12], lycopene [6], lysine [13, 14], propane diol [15], and therapeutic proteins [16]. Some of this work has been adopted by industry and the contribution of metabolic engineering to industrially relevant processes should continue to grow. For example, after studying production of 1,2- and 1,3-propane diol by native organisms, specific enzymes have been transferred to *Escherichia coli* to construct entirely new metabolic pathways that produce these compounds from sugar. Despite initially low titers at approximately 25% of the theoretical yield [17], metabolic engineering and optimization of the pathways has significantly increased titers to the point where Dupont is now commercializing the production of 1,3-propane diol via fermentation using corn starch [18]. Beyond commodity and specialty chemical production, higher value products such as pharmaceutical intermediates can also be produced using metabolic engineering. The construction and optimization of selective trans-(1R, 2R)-indandiol, a key precursor for the AIDS drug Crixivan, has previously been demonstrated [19]. By carefully studying the bioreaction network used in producing this chiral molecule, targeted modifications were implemented to eliminate competing reactions, which resulted in improvement of yield and selectively up to 95% [20].

For many bioprocesses that are the focus of metabolic engineering projects, the competing chemical processes employ nonrenewable fossil resources. These chemical processes often have increased chemical handling and waste that could be reduced by using fermentation technology when existing economic constraints can be met. Almost all fermentation processes are based upon renewable resources as the raw material for making other chemicals. The most common substrates used in these fermentation processes are simple sugars primarily from plant polysaccharides such as cornstarch, which is relatively expensive when compared to chemical feedstocks. Thus, by moving further upstream in the industrial process to the raw material source, metabolic engineering can have an even greater impact on lowering production costs, as shown in Fig. 2. Using metabolic engineering to redesign plants so that they contain a greater percentage of available sugar, are more readily converted into process raw materials, or provide a greater abundance of processing intermediates that can be immediately converted into a final

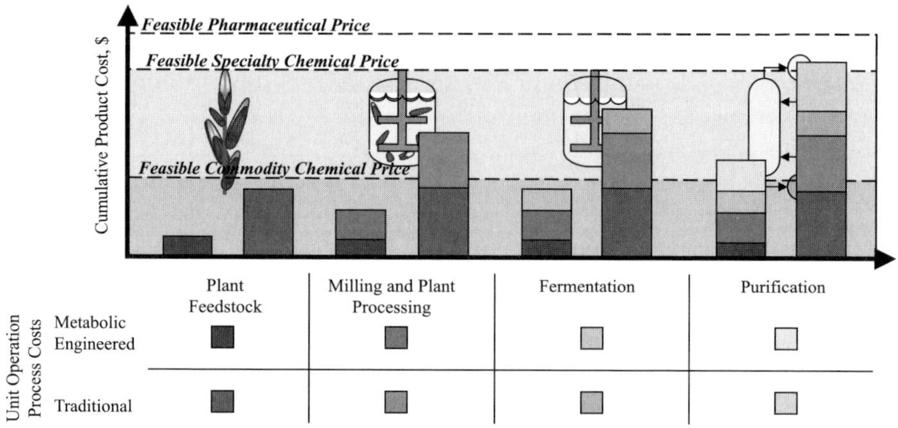

Fig. 2 Economic advantages imparted through metabolic engineering in chemical production. Metabolic engineering can have a large impact on the production of chemicals from agricultural feedstocks. Although the economic advantages that may be potentially imparted by metabolic engineering vary depending upon the exact chemical and process, the figure shows an example of comparisons in which engineering a new feedstock (*hashed bars*) is able to decrease the costs of milling and plant processing, fermentation, and purification relative to processes that have not incorporated metabolic engineering (*vertical bars*). The *dotted lines* represent the relative levels below which certain classes of chemicals become economical

product are all goals of metabolic engineering in agriculture. Further, the potential exists to produce therapeutic proteins in plants, which could eliminate the need for large-scale fermentation or cell-culture facilities and only require purification and formulation processes – a significant decrease in capital expenditures [21–25]. There are many opportunities and challenges for metabolic engineers in this area, including increasing protein production, controlling glycosylation, and altering desirable metabolic pathways.

Beyond its application in industrial and agricultural biotechnology, metabolic engineering principles are becoming increasingly recognized in medicine. Here researchers are often challenged by the integration of data from patients, animal models, and tissue-culture experiments. Systematic approaches afforded by metabolic engineering analyses are becoming more appreciated as ways to integrate diverse data. Data-mining techniques are finding applications in diagnosis [26], as well as helping identify new and important molecules from large data sets. While many of these data sets were initially derived using DNA microarrays, and other high-throughput measurements, metabolite profiling and the in vivo use of isotopic tracers are beginning to emerge as new medical applications of metabolic engineering. In principle animals obey the same laws and constraints as single cells [27] and are amenable to a metabolic engineering analysis. In practice the increased complexity of animals gives rise to special considerations that must

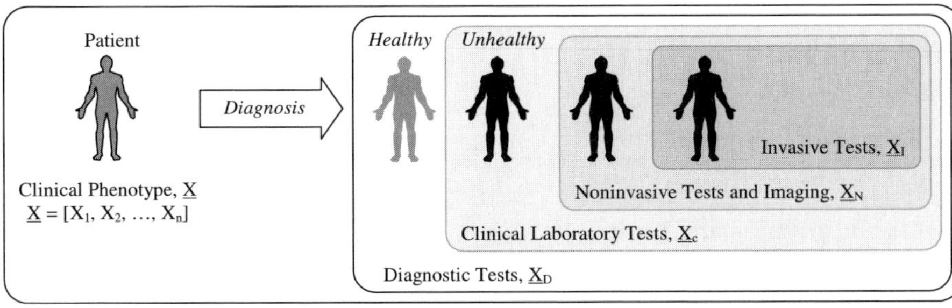

Fig. 3 Incorporation of metabolic engineering tools for clinical diagnosis and treatment. As clinical medicine moves towards an era of personalized healthcare, where each patient's medical status is accurately described by their "clinical phenotype", $\underline{X}$, new diagnostic tests must be developed that can be used to classify patients accurately for increasingly specific treatments based upon measuring elements of $\underline{X}$. The cost of additional tests must be weighed against the probability and expectation that they will return useful information to tailor the patient's therapy. Thus for basic conditions, where few treatments are available, general diagnostic tests, $\underline{X}_D$, where the elements of $\underline{X}_D$ are a subset of X, are conducted. Conversely, for increasingly complex diseases, such as cancer or diabetes, where multiple therapies are available, more tests are warranted, and proceed to add elements from $\underline{X}$ to arrive at new "diagnostic vectors", $\underline{X}_C$, $\underline{X}_N$, $\underline{X}_I$. Metabolic engineering tools can contribute by identifying the most discriminatory variables that can be measured and thereby help reduce costs

be dealt with on a case-by-case basis. Nonetheless, flux measurements and metabolite profiling can be conducted on primary cells isolated from normal, treated, or mutant animals, and promises to enrich our understanding of specific maladies and conditions. Certain disease conditions, such as diabetes mellitus and obesity, are particularly well suited for study by metabolic engineers because they involve sugar metabolism and storage, areas that have been traditionally studied in metabolic engineering. This work may lead to the identification of new surrogate markers for certain diseases, as well as a more quantitative analysis of the in vivo reaction networks that underlie physiology. Advances in this area promise to contribute to personalized medicine by incorporating increasing levels of measurements that can be used to tailor therapies to a person's genetic and metabolic profiles, as described in Fig. 3.

While data-analysis tools represent the foremost application to medicine, metabolic engineering may also provide an expanded framework for gene therapy. Gene therapy, like metabolic engineering, is an attempt to transform a deleterious phenotype into one that is more fit by manipulating specific genes [28]. In developing gene therapy protocols, many of the animal experiments required already follow an algorithm similar to that shown in Fig. 1. Expanding the experimental protocols to include more detailed information about metabolism may be helpful in studying a number of important disease classes including metabolic and neural diseases. Given the complexity of different disease states, metabolic engineering may be used to help identify therapeutic genes that are critical to correcting the genetic component of specific diseases.

3 Metabolic engineering tools

Metabolic engineering relies upon methods that perturb the genome, measure fluxes, and analyze the state of the cell, such that the cell's network architecture can be elucidated and effective targets for genetic manipulation can be identified. An important part of engineering the cell's phenotype is being able to perform the desired genetic perturbations efficiently. Molecular biology provides an array of techniques that can be used to create gene deletions and overexpress genes of interest routinely, making it possible to change the activities of certain enzymes in a desired pathway precisely. This is an essential requirement for metabolic engineering, as the desired change in activity may not be a deletion (no activity) or overexpression with a very strong promoter (order of magnitude change in activity). In some cases a deletion is not possible as the enzyme is required for cell survival. Likewise, strong overexpression can result in deleterious outcomes such as the accumulation of toxic intermediates in a pathway. However, methods that allow the abundance of a necessary enzyme to be reduced or increased by incremental amounts may be able to avoid these problems.

There are several alternatives being developed to control the activity levels of an enzyme precisely. Tuneable promoters attempt to provide a wide range of promoter strengths based on levels of an activator or inhibitor, or simply the promoter sequence. By controlling the copy number of a plasmid, one can control the number of open reading frames in a cell that are available for transcription. In addition, engineering the half-life of RNA transcripts controls the amount of messenger RNA available to be translated into active protein [29].

Several advances in applied molecular biology are allowing metabolic engineers to take advantage of nature's inherent biodiversity by using combinatorial techniques to more efficiently sample and select beneficial traits from cellular systems. High-efficiency transformations allow libraries of 10^9 genetic variants to be generated. Transposon mutagenesis enables a high-throughput form of mutagenesis where there is only one mutation (resulting from the insertion of a stabilized transposable element) introduced per cell [30]. The location of the insertion can be routinely determined by sequencing from the transposable element. This technique is a large improvement over classical mutagenesis methods where multiple mutation sites were common and the site of a mutation was more difficult to locate. Gene shuffling and directed evolution are other methods that allow not only changes in the expression levels of an enzyme but also can be used to engineer the specificity and alter post-translational regulation [31].

Once the network has been perturbed, we must understand how it responds to the perturbation. This is done by comparing the metabolic phenotype of the perturbed network to the unperturbed control network. Methods that enable measurement of metabolic fluxes have been developed to give information on the metabolic phenotype [1]. These high-throughput methods are used to assay the in vivo levels of many metabolites easily and thereby measure multiple fluxes as they appear in the system. Determining the fluxes often requires the measurements to be made at a metabolic steady state and most commonly incorporates metabolite labeling. ^{13}C-labeling is often chosen because virtually all molecules of interest in the network contain carbon, but many other isotopes are available to tailor an experiment. As the labeled substrate proceeds through the metabolic network, the pools of metabolites that are downstream from the substrate become labeled. At steady state the fraction of labeled substrate in a given pool can be used to calculate the flux through that pathway.

The fate of individual carbon atoms can be tracked using *positional isotopomers*. In general for an organic molecule composed of n carbon atoms, there are 2^n possible isotopomers. These isotopomers can be observed by gas chromatography-mass spectrometry (GC-MS) or nuclear magnetic resonance (NMR) spectroscopy. The intracellular fluxes determine the distribution of the positional isotopomers through the various pathways. For example, lysine can be produced from oxaloacetate and pyruvate via two different pathways. In one pathway, the six carbons contained in lysine are derived from the four carbons of oxaloacetate and two terminal carbons of pyruvate; conversely, in the other pathway the carbons are derived from three terminal carbon atoms from oxaloacetate along with all three of pyruvate's carbon atoms. Thus using different isotopic-labeling patterns within the substrate molecules will result in differentially labeled lysine molecules, the abundance of which depends upon the fluxes within the two pathways. By measuring the distribution of lysine isotopomers, the quantitative fluxes can be calculated [32, 33]. It should be noted that it is important to close the isotopic material balance to help ensure consistency among the measurements and to provide reliable comparisons between experiments. To measure steady-state metabolite levels, chemostats are often a convenient method for culturing cells. Once a chemostat has reached steady state, the flux of extracellular metabolites into or out of the cells can be calculated measuring the difference in concentration of the metabolite between the feed and exit stream. This measurement divided by the time constant for the chemostat gives the specific uptake or release of a given metabolite by the culture.

In the case where the flux through a linear pathway is of interest, isotopomer methods are insufficient. Without splitting the carbon backbone, the

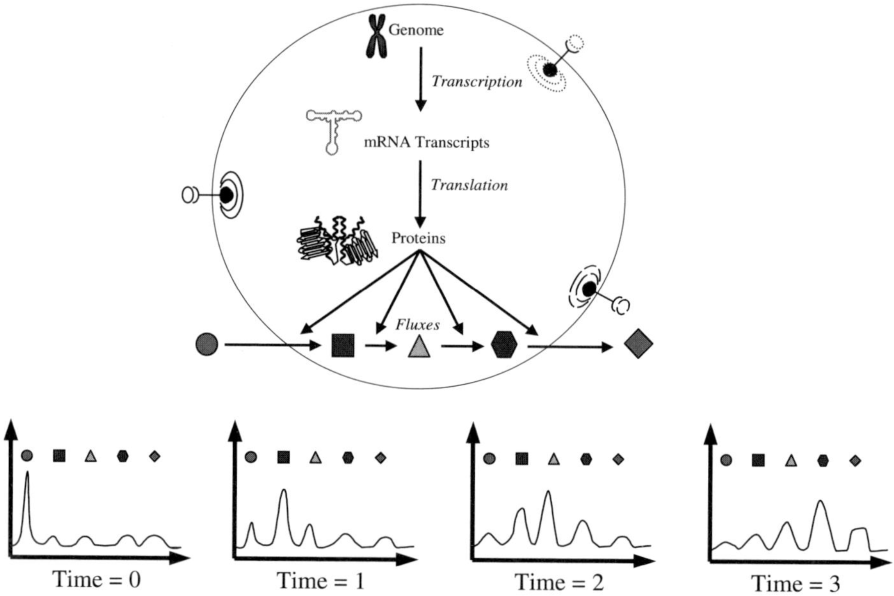

Fig. 4 Determination of flux through a linear pathway. The figure illustrates how one may determine the flux through a linear pathway by treating the cells with a pulse of labeled substrate under steady-state conditions. In this figure, the concentration of each metabolite, designated by a different shape, is determined over time following the introduction of the labeled substrate

levels of labeled metabolites will remain the same in a linear pathway. In these situations, transient isotope feeds have been used in a metabolic steady state to reveal the flux in these linear pathways. Specifically, a pulse of radioactive ^{14}C substrate is taken up by the cell and the amount of radioactive isotope in each metabolite pool is then measured in time as shown in Fig. 4. The rate of accumulation and depletion in each metabolite pool can be used to estimate the flux through the pathway [7].

Given that we now have methods to measure metabolite pools in specifically controlled conditions, next we want to calculate the carbon fluxes throughout the cell. The intracellular fluxes can only be partially estimated from external metabolite uptake or release. The problem can be posed in matrix notation, as shown in Eq. 1 where $\boldsymbol{r}$ is a vector of the specific uptake or secretion rates of extracellular metabolites (mol/s/cell), $\boldsymbol{G}$ is the matrix containing stoichiometric coefficients for the metabolic reactions, and $\boldsymbol{v}$ is a vector of reaction rates for the biochemical system (mol/s/cell). In $\boldsymbol{G}$, rows represent reactions and columns are the metabolites involved in each reaction.

$$\boldsymbol{r} = \boldsymbol{G}^{\mathrm{T}} \boldsymbol{v} . \tag{1}$$

In some situations, such as those harboring parallel, redundant, or reversible pathways, $\boldsymbol{G}$ is not invertible, making it impossible to solve for the fluxes. In these cases, NMR/GC-MS methods can be used to measure the levels of labeled intracellular metabolites. The raw ^{13}C-NMR, GC-MS measurements can be used to calculate the fluxes of carbon through the cell. As mentioned previously, the distribution of labeled metabolites in the cell determine the intracellular fluxes. Given the measurements, a linear set of relationships, subject to stoichiometric constraints can be formulated. Depending on the number of observables, the system may be overdetermined (more measurements than fluxes) or underdetermined (more fluxes than measurements). For an overdetermined system, the redundant measurements can be used to add statistical information to the measurements and check for gross errors [34]. In the situation of an underdetermined system, a linear programming problem must be formulated where an objective function is optimized subject to the metabolite balance constraints. The exact form of the objective function may vary, but among the most commonly reported are specific growth rate, cellular energetics, or substrate utilization. Constraints other than the metabolite balance have been successfully used to improve the linear optimization by restricting the in silico solution space to more closely represent the possible fluxes in a cell [35]. These constraints are often based on enzyme capacity and the thermodynamics associated with reaction directionality. Although the so determined "optimized fluxes" are not necessarily equal to the actual fluxes, they have nevertheless been used as flux surrogates in several cases [36].

The methods and models used to calculate the intracellular fluxes can now be directed toward determining how to manipulate the cell to achieve the desired phenotype. After measuring the fluxes through the metabolic network, it is necessary to identify the pathways and enzymes that will most drastically improve the phenotype. Metabolic control analysis (MCA) provides a framework to help understand how flux control is distributed in a bioreaction network. Finding enzyme (gene) targets having the greatest influence on a product rate can be difficult because a rate-limiting step is often not found in biological networks. Instead the limitations are spread over many enzymes in the network. The flux control coefficient (FCC) of an enzyme is defined as the relative effect of modulating the amount of an enzyme on the flux through the desired pathway. Equation 2 shows the flux control coefficient C_i^J of an enzyme E_i on the flux J.

$$C_i^J = \frac{\mathrm{d}J}{\mathrm{d}E_i}\left(\frac{E_i}{J}\right) \qquad (2)$$

The FCC is essentially a sensitivity coefficient of the flux with respect to various enzymes. An important property of the FCC is that summation of all the

FCCs affecting a particular flux must equal unity (Eq. 3).

$$\sum_i C_i^J = 1 \qquad (3)$$

An FCC that approached unity would imply a rate-limiting enzyme. FCCs in a linear pathway will all be positive and less than 1, while a competing pathway may have a negative FCC. For an enzyme with a low FCC, a manyfold increase in the activity of an enzyme may only change the final product marginally. In practice, a variety of experiments must be performed to determine where the flux control is located in the network [37].

Despite the large amount of effort in determining FCCs, the result is a comprehensive understanding of which enzymes in the network should be targeted and how much of an improvement can be expected for a given target (based on the magnitude of the FCC). In general, MCA is useful for conceptualizing kinetic limitations in bioreaction networks, as well as analyzing small well-defined pathways. When analyzing larger systems, the group flux control coefficient (gFCC) is a more succinct way to evaluate what is important for the flux of interest. The gFCC allows the grouping of branches of metabolism together (for example one group might be the pentose phosphate pathway and another may be the citric acid cycle) to identify which regions of metabolism are important to controlling the flux of interest. MCA, while experimentally intensive, provides a framework for elucidating the control of a network [38].

4 New contributions to metabolic engineering

Progress in related areas of biology has provided new tools for metabolic engineers. While the mathematical analyses and use of isotopic tracers developed previously are still important, tools from other areas are being incorporated into the metabolic engineer's repertoire [39]. Similar to metabolite profiling, transcription profiling using DNA microarrays can provide information about the level of gene activation on a genome-wide basis. While it may seem intuitive that genes encoding enzymes that catalyze specific reactions are necessarily the targets for control, the actual situation is often much more complicated. Repressors, enhancers, and even epigenetic events can influence gene regulation and are often influenced by extracellular signals. In addition, enzyme activity can be modulated by post-translational modification that may result from the stimulation of other genes that are not intuitively obvious. Thus, transcription monitoring has an essential role in upgrading the information content derived from flux analysis and linking it to the genes that ultimately control cellular physiology. DNA microarrays have also been employed by the metabolic engineering community to identify the genes responsible for specific, selected traits. In circumstances where a selective pressure can be applied, such as growth in the presence of an inhibitory/toxic compound or on a new substrate, to organisms transformed with a plasmid library, fit organisms that survive the selection process can be immediately "sequenced" after labeling their purified plasmids and hybridizing them to a DNA microarray [40].

High-throughput methods of gene manipulation also provide a way of rapidly screening for new metabolic performance. In the case of bacteria, the use of transposable elements has enabled researchers to generate large libraries of knockout mutants quickly, which can be subsequently screened for greater titers or improved flux performance. This technique complements the usual method of directed gene knockout via homologous recombination. In a similar manner for mammalian cells, genes identified from microarray experiments or flux balance analysis can be specifically silenced using RNA interference [41]. In addition, large-scale screening experiments can also be employed using this method [42] and provide a technique for the generation of null phenotypes that is easy to use and was previously unavailable.

Metabolite profiling is another technique developed by metabolic engineers that is quickly gaining acceptance in a wide variety of applications. Similar to transcriptional profiling, measuring the abundance of cellular metabolites provides a broad glimpse of the metabolic cellular state. However, unlike previously mentioned isotopic-labeling methods, metabolic profiling does not attempt to establish the intracellular flux, making this experimentally more convenient. Nevertheless, it may be that the metabolite profiles provide enough similar information such that, when combined with protein and transcript profiles, a fairly complete picture of the cell is obtained that can be used to solve more complex systemic problems.

One of the problems currently facing researchers is how to integrate the large, diverse data sets that are generated from high-throughput technologies. While traditional modeling approaches used in metabolic engineering, such as flux balance analysis, cannot readily accommodate different data types, metabolic control theory could in principle. However, in practice it is not always possible to control genetic variables adequately to determine metabolic control coefficients. Instead, new analysis techniques will need to be employed. Statistical modeling, such as partial least squares [43], has the ability to relate different data matrices generated via high-throughput experimental procedures immediately and thereby upgrade the information content of the data.

5 Conclusion

In the past, determining metabolic fluxes within an organism was a substantial undertaking. Besides obtaining specifically labeled molecules, which could be challenging, and achieving a steady state within a continuous reactor, this work was often additionally complicated by the lack of information regarding an organism's metabolic pathways. As increasing numbers of organisms are fully sequenced and more thoroughly investigated, many of the previous constraints associated with network definition are being removed and indeed new hypotheses can be constructed from the sequence information alone.

The expansion in our knowledge base has been accompanied by improved experimental technologies. Isotopic tracer experiments are being implemented more routinely, and metabolite profiling enables researchers to detect hundreds of metabolites in a single experiment. Other high-throughput technologies, such as DNA microarrays and proteomics tools, have allowed researchers to measure more cell parameters with substantially less effort. This has resulted in a shift from localized studies to *systems biology* investigations. As new experimental techniques are expanding the number of variables that can be incorporated into the analysis, enormous data sets are being generated. Metabolic engineering is well suited to utilize this wealth of data and provides a rational framework for incorporating these new experimental methods.

A new paradigm based on combinatorial searches is emerging to exploit metabolic engineering principles. The ability to create large libraries of microorganisms that over- or underexpress specific genes, and efficiently screen or select for desirable properties, is enabling a new high-throughput approach to metabolic engineering. New technologies that enable massively parallel screening for a wide variety of non-growth-associated phenotypes will be critical to these developments. Strategies to search the combinatorial space have as their foundation the previous metabolic engineering paradigm that often dealt with information-deficient systems and limited experimental tools, and are therefore focused on directed manipulation of specific genes within a cell. The new paradigm that is developing for metabolic engineering takes advantage of tools to create numerous mutations, select, and then importantly *identify* the causative changes in combinatorial experiments. When combined with metabolic engineering's framework of analysis, this creates a very powerful strategy for searching the phenotype space available to an organism, and quickly evolving changes that improve the desired qualities.

Implementation of these emerging tools creates an opportunity to advance metabolic engineering into new areas of application. This opportunity comes at a critical time as the economic potential of biotechnology is increasingly realized throughout industrial innovation. Further use of metabolic engineering in medicine, agriculture, and bioprocessing can complement other technical achievements in those fields and hopefully contribute to overcoming scientific challenges in these areas.

Acknowledgements We would like to thank the National Science Foundation for their funding through NSF Grant: BES-0331364, as well as the Singapore-MIT Alliance for additional funding.

References

1. Stephanopoulos G (1999) Metabolic fluxes and metabolic engineering. Metab Eng 1:1–11
2. Stephanopoulos G, Vallino JJ (1991) Network rigidity and metabolic engineering in metabolite overproduction. Science 252:1675–1681
3. Bailey JE (1991) Toward a Science of Metabolic Engineering. Science 252:1668–1675
4. Sudesh K, Taguchi K, Doi Y (2002) Effect of increased PHA synthase activity on polyhydroxyalkanoates biosynthesis in Synechocystis sp PCC 6803. Int J Bio Macromol 30
5. Niederberger P, Prasad R, Miozzari G, Kacser H (1992) A strategy for increasing an in vivo flux by genetic manipulations. The tryptophan system of yeast. Biochem J 287:473–479
6. Farmer WR, Liao JC (2000) Improving lycopene production in Escherichia coli by engineering metabolic control. Nat Biotechnol 18:533–537
7. Lu JL, Liao TC (1997) Metabolic engineering and control analysis for production of aromatics: Role of transaldolase. Biotechnol Bioeng 53:132–138
8. Ostergaard S, Olsson L, Johnston M, Nielsen J (2000) Increasing galactose consumption by Saccharomyces cerevisiae through metabolic engineering of the GAL gene regulatory network. Nat Biotechnol 18:1283–1286
9. Aiba S, Matsuoka M (1979) Identification of metabolic model: Citrate production from glucose by Candida lipolytica. Biotechnol Bioeng 21:1373–1386
10. Stafford D, Yanagimachi K, Stephanopoulos G (2001) Metabolic engineering of indene bioconversion in Rhodococcus sp. Adv Biochem Eng Biotechnol 73:85–101
11. Ohta K, Beall DS, Mejia JP, Shanmugam KT, Ingram LO (1991) Metabolic Engineering of Klebsiella-Oxytoca M5a1 for Ethanol-Production from Xylose, Glucose. Appl Env Microbiol 57:2810–2815
12. van Maris AJA, Konings WN, van Dijken JP, Pronk JT (2004) Microbial export of lactic and 3-hydroxypropanoic acid: implications for industrial fermentation processes. Metab Eng 6:245–255
13. Koffas MAG, Jung GY, Aon JC, Stephanopoulos G (2002) Effect of pyruvate carboxylase overexpression on the physiology of Corynebacterium glutamicum. Appl Env Microbiol 68:5422–5428
14. Koffas MAG, Jung GY, Stephanopoulos G (2003) Engineering metabolism and product formation in Corynebacterium glutamicum by coordinated gene overexpression. Metab Eng 5:32–41
15. Tong IT, Liao HH, Cameron DC (1991) 1,3-Propanediol production by Escherichiacoli expressing genes from the klebsiella-pneumoniae-dha regulon. Appl Env Microbiol 57:3541–3546
16. Vives J, Juanola S, Cairo JJ, Godia F (2003) Metabolic engineering of apoptosis in cultured animal cells: implications for the biotechnology industry. Metab Eng 5:124–132
17. Cameron DC, Altaras NE, Hoffman ML, Shaw AJ (1998) Metabolic engineering of propanediol pathways. Biotechnol Progr 14:116–125

18. Danner H, Braun R (1999) Biotechnology for the production of commodity chemicals from biomass. Chem Soc Rev 28:395–405
19. Buckland BC et al. (1999) Microbial conversion of indene to indandiol: a key intermediate in the synthesis of CRIXIVAN. Metab Eng 1:63–74
20. Stafford DE et al. (2002) Optimizing bioconversion pathways through systems analysis and metabolic engineering. Proc Natl Acad Sci USA 99:1801–1806
21. Hood EE, Woodard SL, Horn ME (2002) Monoclonal antibody manufacturing in transgenic plants – myths and realities. Curr Opin Biotechnol 13:630–635
22. Larrick J, Yu L, Naftzger C, Jaiswal S, Wyco K (2002) In: Hood E, Howard J (eds.) Plants as factories for protein production. Kluwer Academic, Boston. pp. 79–101
23. Morrow KJ (2002) Economics of antibody production – Various options available for large-scale bioprocessing. Genet Eng News 22:1–39
24. Nikolov Z, Hammes D (2002) In: Hood E, Howard J (eds) Plants as factories for protein production. Kluwer Academic, Boston. pp. 159–174
25. Thiel KA (2004) Biomanufacturing, from bust to boom...to bubble? Nat Biotechnol 22:1365–1372
26. Stephanopoulos G (2000) Bioinformatics, metabolic engineering. Metabol Eng 2:157–158
27. Lavoisier AL, DeLaplace PS (1994) Memoir on heat. Obes Res 2:189–203
28. Wang F, Raab RM, Washabaugh MW, Buckland BC (2000) Gene therapy, metabolic engineering. Metab Eng 2:126–139
29. Keasling JD (1999) Gene-expression tools for the metabolic engineering of bacteria. Trends Biotechnol 17:452–460
30. Goryshin IY, Jendrisak J, Hoffman LM, Meis R, Reznikoff WS (2000) Insertional transposon mutagenesis by electroporation of released Tn5 transposition complexes. Nat Biotechnol 18:97–100
31. Tobin MB, Gustafsson C, Huisman GW (2000) Directed evolution: the 'rational' basis for 'irrational' design. Curr Opin Struc Biol 10:421–427
32. Park SM, Klapa MI, Sinskey AJ, Stephanopoulos G (1999) Metabolite and isotopomer balancing in the analysis of metabolic cycles: II. Applications. Biotechnol Bioeng 62:392–401
33. Klapa MI, Park SM, Sinskey AJ, Stephanopoulos G (1999) Metabolite and isotopomer balancing in the analysis of metabolic cycles: I. Theory. Biotechnol Bioeng 62:375–391
34. Klapa MI, Aon JC, Stephanopoulos G (2003) Systematic quantification of complex metabolic flux networks using stable isotopes and mass spectrometry. Eur J Biochem 270:3525–3542
35. Price ND, Papin JA, Schilling CH, Palsson BO (2003) Genome-scale microbial in silico models: the constraints-based approach. Trends Biotechnol 21:162–169
36. Edwards JS, Ibarra RU, Palsson BO (2001) In silico predictions of Escherichia coli metabolic capabilities are consistent with experimental data. Nat Biotechnol 19:125–130
37. Fell D (1997) Understanding the control of metabolism. Portland, Brookfield, VT
38. Stephanopoulos G, Aristidou AA, Nielsen J (1998) Metabolic engineering: principles, methodologies. Academic, San Diego
39. Nielsen J (2003) It is all about metabolic fluxes. J Bacteriol 185:7031–7035
40. Gill RT, Wildt S, Yang YT, Ziesman S, Stephanopoulos G (2002) Genome wide screening for trait conferring genes using DNA micro-arrays. P Natl Acad Sci USA 99:7033

41. Raab RM, Stephanopoulos G(2004) Dynamics of gene silencing by RNA interference. Biotechnol Bioeng 88:121–132
42. Ashrafi K et al. (2003) Genome-wide RNAi analysis of Caenorhabditis elegans fat regulatory genes. Nature 421:268–272
43. Chan C, Hwang D, Stephanopoulos GN, Yarmush ML, Stephanopoulos G (2003) Application of multivariate analysis to optimize function of cultured hepatocytes. Biotechnol Progr 19:580–598

Adv Biochem Engin/Biotechnol (2005) 100: 19–51
DOI 10.1007/b136410

Published online: 5 July 2005

Microbial Isoprenoid Production: An Example of Green Chemistry through Metabolic Engineering

Jérôme Maury[1] · Mohammad A. Asadollahi[1] · Kasper Møller[1] · Anthony Clark[2] · Jens Nielsen[1] (✉)

[1]Center for Microbial Biotechnology, BioCentrum-DTU, Building 223, Technical University of Denmark, 2800 Kgs. Lyngby, Denmark
jn@biocentrum.dtu.dk

[2]Firmenich, Route des Jeunes 1, 1211 Genève 8, Switzerland

Abstract Saving energy, cost efficiency, producing less waste, improving the biodegradability of products, potential for producing novel and complex molecules with improved properties, and reducing the dependency on fossil fuels as raw materials are the main advantages of using biotechnological processes to produce chemicals. Such processes are often referred to as green chemistry or white biotechnology. Metabolic engineering, which permits the rational design of cell factories using directed genetic modifications, is an indispensable strategy for expanding green chemistry. In this chapter, the benefits of using metabolic engineering approaches for the development of green chemistry are illustrated by the recent advances in microbial production of isoprenoids, a diverse and important group of natural compounds with numerous existing and potential commercial applications. Accumulated knowledge on the metabolic pathways leading to the synthesis of the principal precursors of isoprenoids is reviewed, and recent investigations into isoprenoid production using engineered cell factories are described.

Keywords Green chemistry · Metabolic engineering · Cell factories · Isoprenoids

Abbreviations

ATP	Adenosine triphosphate
CDP-ME	4-diphosphocytidyl-2C-methyl-D-erythritol
CDP-ME2P	2-phospho-4-diphosphocytidyl-2C-methyl-D-erythritol
CMP	Cytidine monophosphate
CTP	Cytidine triphosphate
CoA	Coenzyme A
DMAPP	Dimethylallyl diphosphate
DXP	1-deoxy-D-xylulose 5-phosphate
ERAD	Endoplasmic reticulum associated degradation
FOH	Farnesol
FPP	Farnesyl diphosphate
GAP	D-glyceraldehyde 3-phosphate
GGPP	Geranylgeranyl diphosphate
GMO	Genetically modified organism
GPP	Geranyl diphosphate
HMBPP	1-hydroxy-2-methyl-2-(*E*)-butenyl 4-diphosphate
HMG-CoA	3-hydroxy-3-methylglutaryl coenzyme A
IPP	Isopentenyl diphosphate
MECDP	2-C-methyl-D-erythritol 2,4-cyclodiphosphate
MEP	2-methylerythritol 4-phosphate
MCA	Metabolic control analysis
MFA	Metabolic flux analysis
mRNA	Messenger ribonucleic acid
NADP	Nicotinamide adenine dinucleotide phosphate
PEP	Phosphoenolpyruvate
RNA	Ribonucleic acid
TPP	Thiamine diphosphate
tRNA	Transfer ribonucleic acid

1 Introduction

Cell factories are extensively applied to produce many specific molecules that are used as pharmaceuticals, fine chemicals, fuels, materials and food ingredients. There is much focus on the production of recombinant proteins, with a current market value exceeding 40 billion US$, but the market for small molecules is larger and is expected to grow faster in the future. The main driving force behind this growth is directed genetic modifications of cell factories—an approach referred to as *metabolic engineering*. Metabolic engineering enables the development of novel and efficient bioprocesses that are environmentally friendly [1–4], and makes use of cell factories to produce novel compounds that are difficult to produce by organic chemical synthesis. Many top-selling drugs are natural products [5]—they accounted for approximately 40% of the top twenty drugs in 1997 [6]—and it is anticipated that natural products will provide an increasing number of new drugs in the future. Therefore, classical chemical synthesis is increasingly being replaced by biotech processes; indeed the Department of Energy in the USA has predicted that the market size of biotech-derived small molecules will exceed 100 billion US$ in 2010 and 400 billion US$ in 2030, and will then represent about 50% of the market for organic molecules. Another report from McKinsey and Company [7] predicts that up to 20% of all organic chemicals will be produced via biotechnological routes by 2010 (Fig. 1). The use of biotechnology to produce chemicals is often referred to as **green chemistry**; in Europe the term *white biotechnology* is often used (Table 1). The key drivers for this development towards green chemistry are:

- Biotech processes can in many cases be designed as integrated processes with small waste streams, and they are more energy efficient and more resource efficient than classical chemical processes.
- Biotech products are biodegradable and so they represent an improved lifecycle for the products.
- Biotech offers the potential to produce chemicals with a huge diversity, achieving novel structures that are almost impossible to obtain using traditional organic chemical synthesis.

During the development of novel bioprocesses (or the improvement of existing bioprocesses), the value added element is primarily in the design of efficient cell factories. There are several large research groups and companies focusing on the development of cell factories for novel and/or improved bioprocesses worldwide. Traditionally, biotech processes have been developed based on screening for a microorganism with interesting properties (for ex-

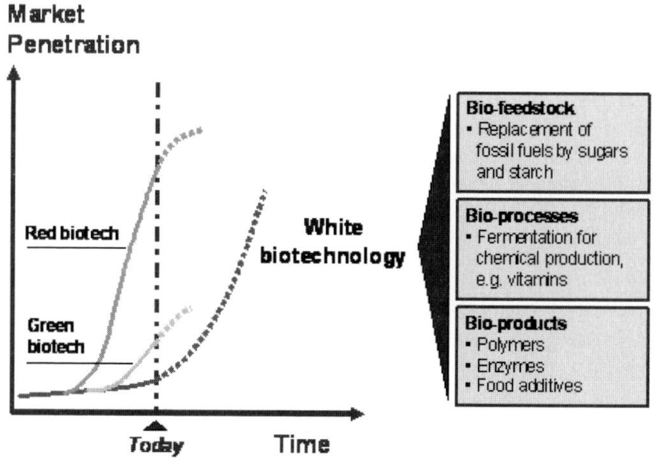

Fig. 1 Predicted market penetration of white biotechnology, which is also referred to as "the application of nature's toolset to industrial production" [7]. The figure is adapted from [7]

Table 1 Some definitions of different applications of biotechnology

Term	Definition
Red Biotechnology	Production of pharmaceutical proteins using biotechnology, i.e. using different cell factories. Generally the products are high-value added products and they are produced in relatively small volumes.
Green Biotechnology	The use of plants in biotechnology, e.g. use of GMO plants for production of polymers.
White Biotechnology/ Green Chemistry	The use of biotechnology in industrial processes, therefore also often referred to as industrial biotechnology. More specifically these terms encompass production of bulk and fine chemicals, e.g. amino acids, vitamins, antibiotics, enzymes, organic acids, polymers and other chemicals. Basically green chemistry, white biotechnology and industrial biotechnology describe the same thing.

ample, it produces an interesting compound), whereas in recent years there has been a paradigm shift towards the use of a few well-chosen cell factories. Good examples of this are: 1) the use of a few selected microorganisms to produce a wide range of different enzymes (the Danish company Novozymes has expressed a large number of different enzymes in the filamentous fungus *Aspergillus oryzae*), 2) the use of the penicillin-producing fungus *Penicillium chrysogenum* by the Dutch company DSM for the production of adipoyl-7-aminodeacetoxycephalosporanic acid (adipoyl-7-ADCA) [8], a precursor for the production of semi-synthetic cephalosporins, and 3) the production of the chemical 1,3-propanediol by the American company Dupont by a recombinant *Escherichia coli*, an organism that is already used for the production of many other chemicals, such as phenylalanine. There are several drivers for this development, including:

- Scale-up of bioprocesses can be intensified; when a cell factory has already been used for the production of different products there is extensive empirical knowledge on how a new process based on this cell factory can be scaled-up.
- Fundamental research on the cell factory pays off, as it may impact several different processes. Furthermore, deeper insight into the function of the cell factory is gained through fundamental research, and this enables even wider use of the cell factory for industrial production.
- It may be easier to obtain process (and product) approval when cell factories that are already well implemented are applied.

In the following, the move towards a wider use of green chemistry is exemplified by the recent endeavors to develop suitable cell factories capable of accumulating significant amounts of isoprenoids, a widespread group of natural compounds with numerous existing and potential applications.

2 Microbial Isoprenoid Production

2.1 Isoprenoids

Isoprenoids (also referred to as terpenoids) are a diverse group of natural compounds with more than 23 000 identified compounds [9]; most of them are found in plants as constituents of essential oils [10]. Isoprenoids are derived from five-carbon isoprene units (2-methyl-1,3-butadiene) and the combination of isoprene units leads to the formation of different isoprenoids. Based on the 'isoprene rule' that was first recognized in 1887 by Wallach [11] and that was later, in 1953, extended into the 'biogenetic isoprene rule' by Ruzicka [12], isoprenoids can be divided into different groups depending on the number of isoprene units in their carbon skeleton (Table 2).

The universal biological precursor for all isoprenoids is isopentenyl diphosphate (IPP) (Fig. 2). Since the 1960s, when Bloch and Lynen discovered the mevalonate pathway for cholesterol synthesis [13, 14] and until recently, IPP was assumed to be synthesized through the mevalonate-dependent pathway in all living organisms. However, in the 1990s, the existence of an alternative pathway, called the 2-methylerythritol 4-phosphate (MEP) pathway, was demonstrated in bacteria, green algae, and higher plants [15–18].

Isoprenoids are functionally important in many different parts of cell metabolism such as photosynthesis (carotenoids, chlorophylls, plastoquinone), respiration (ubiquinone), hormonal regulation of metabolism (sterols), regulation of growth and development (gibberellic acid, abscisic

Table 2 Classification of isoprenoids based on the number of isoprene units

Class	Isoprene units	Carbon atoms	Formula
Monoterpenoids	2	10	$C_{10}H_{16}$
Sesquiterpenoids	3	15	$C_{15}H_{24}$
Diterpenoids	4	20	$C_{20}H_{32}$
Sesterterpenoids	5	25	$C_{25}H_{40}$
Triterpenoids	6	30	$C_{30}H_{48}$
Tetraterpenoids	8	40	$C_{40}H_{64}$
Polyterpenoids	> 8	> 40	$(C_5H_8)_n$

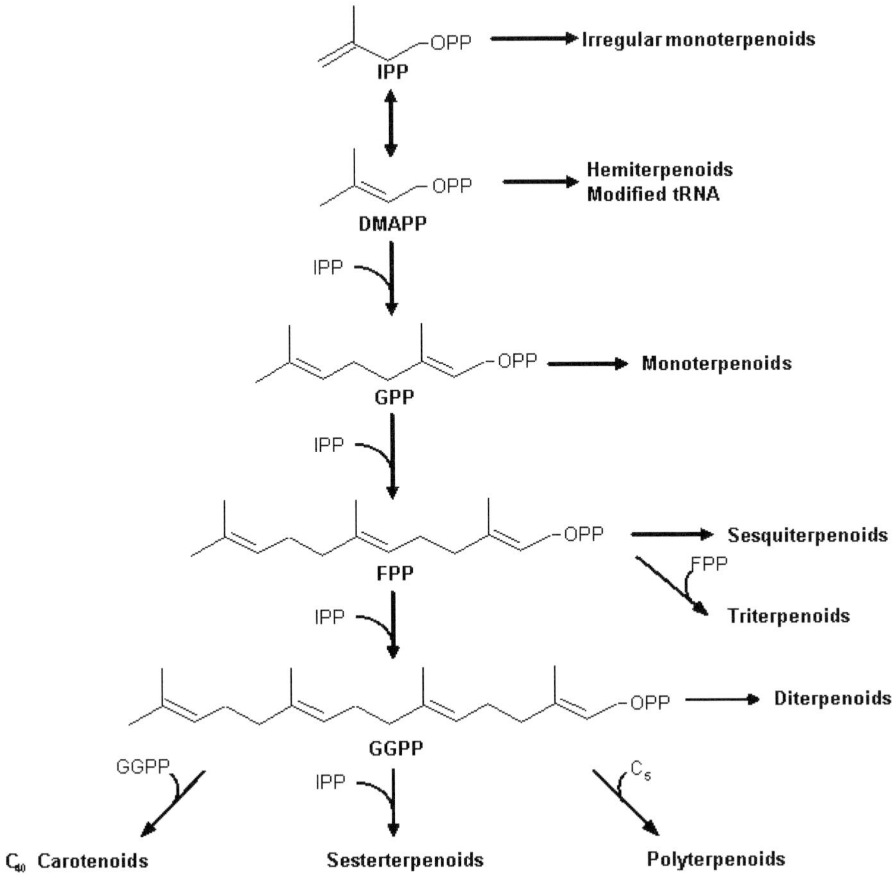

Fig. 2 The different classes of isoprenoids and their precursors DMAPP: dimethylallyl diphosphate, IPP: isopentenyl diphosphate, GPP: geranyl diphosphate, FPP: farnesyl diphosphate, GGPP: geranylgeranyl diphosphate

acid, brassinosteroids, cytokinins, prenylated proteins), defense against pathogen attack, intracellular signal transduction (Ras proteins), vesicular transport within the cell (Rab proteins) as well as defining membrane structures (sterols, dolichols, carotenoids) [9, 19]. Many isoprenoids also have considerable medical and commercial interest as flavors, fragrances (such as limonene, menthol, camphor), food colorants (carotenoids) or pharmaceuticals (such as bisabolol, artemisinin, lycopene, taxol). In Table 3, some examples of isoprenoids and their corresponding biological functions or commercial applications are listed.

Isoprenoids are widely present in plant tissues, and extraction from plants has been the traditional option for the large-scale production of these compounds. However, in many cases this method is neither feasible nor eco-

Table 3 Biological activities or commercial applications of typical isoprenoids

Class	Biological activities[a]	Commercial applications[a]	Examples
Monoterpenoids	Signal molecules, e.g. as defence mechanism against pathogens	Flavors, fragrances, cleaning products, anticancer agents, antimicrobial agents	Limonene, menthol, camphor
Sesquiterpenoids	Antibiotic, antitumor, antiviral, immuno-suppressive, and hormonal activities	Flavors, fragrances, potential pharmaceuticals	Juvenile hormone, nootkatone, artemisinin
Diterpenoids	Hormonal activities, antitumor properties	Anticancer agents	Gibberellins, phytol, taxol
Sesterterpenoids	Cytostatic activities	None as yet	Haslenes
Triterpenoids	Membrane components	Biological markers	Sterols, hopanoids
Tetraterpenoids	Antioxidants, photosynthetic components, pigments, and nutritional elements	Food additives (colorants, antioxidants), anticancer agents	Lycopene, β-carotene
Polyterpenoids	N-linked protein glycosylation, side chains of ubiquinones	Rubber	Dolichols, prenols/q

[a] Biological functions and commercial applications are selected examples.

nomical. Among the drawbacks in using plants as a source for isoprenoid production are influence of geographical location and weather on the composition and concentration of isoprenoids in the plant tissues, low concentration and poor yields for the recovery of isoprenoids from plants, and the high costs associated with extraction and purification. Koepp et al. [20] reported extraction of only 1 mg of 85% taxadiene from 750 kg of bark powder from Pacific yew (*Taxus brevifolia*) after an extensive isolation and purification process. Chemical synthesis of isoprenoids has also been reported [21–23], and currently most of the industrially interesting carotenoids are produced via chemical synthesis [24]. However, because of the complex structures of isoprenoids, chemical synthesis, involving many steps, is difficult. Side reactions, unwanted side products, and low yield are other disadvantages. In vitro enzymatic production of isoprenoids through the action of plant isoprenoid synthases is also impractical due to the dependency on the expensive precursors, as well as poor in vitro conversion.

Microbial production of chemicals is an accepted environmentally friendly method that may lead to the production of a large amount of high-value isoprenoids from simple and cheap carbon sources. Engineered microorganisms would also enable production of unusual and novel isoprenoids with excellent biological and commercial applications.

Directed manipulation of cell factories using genetic engineering techniques requires detailed information about the metabolic pathways and enzymes involved in the biosynthesis of the desired product(s) and also an understanding of the mechanisms by which the flux through the pathway is controlled. One of the major obstacles to the commercial production of isoprenoids by cell factories is the limited supply of precursors. Replenishing the intracellular pool of precursors will need deregulation of pathways in order to improve the flux towards the biosynthesis of isoprenoid precursors. Therefore, before dealing with the investigations conducted in order to produce enhanced strains capable of isoprenoid production, we will discuss the metabolic pathways for isoprenoid biosynthesis, their enzymes and genes and also the regulatory network of pathways.

2.2 The Mevalonate Pathway of *Saccharomyces cerevisiae*

Due to the involvement of isoprenoids in a variety of physiologically- and medically-important processes, the sterol biosynthetic pathway or mevalonate pathway has been intensively studied in eukaryotes. Principal end products of the mevalonate pathway are sterols, such as cholesterol in animal cells and ergosterol in fungi, which are important regulators of membrane permeability and fluidity [25, 26]. In addition to sterols, the mevalonate pathway provides intermediates for the synthesis of a number of other essential cellular constituents like hemes, quinones, dolichols or isoprenylated proteins, which are all derived from the early part of the pathway, prior to the formation of the first cyclic sterol molecule [27]. Thus, the mevalonate pathway can be considered to consist of two distinct parts: an early isoprenoid section of the pathway, common to many branches and ending with the formation of farnesyl diphosphate (FPP), and a late part of the pathway mainly dedicated to ergosterol biosynthesis in *S. cerevisiae* (Fig. 3). This partition of the pathway is also reflected in the oxygen requirements of some enzymatic steps in the second part of the pathway, while this constraint does not exist for the first part of the pathway (Fig. 3). As the early steps of the mevalonate pathway generate precursors for isoprenoid production, the next paragraphs will focus on the enzymes catalyzing these steps, with emphasis on the key regulatory points of the pathway.

The first reaction of the mevalonate pathway is the synthesis of acetoacetyl-CoA from two molecules of acetyl-CoA, catalyzed by the acetoacetyl-CoA thiolase which is encoded by *ERG10* (Fig. 3). *S. cerevisiae* contains two forms of the enzyme, which have different subcellular locations (the cytosol and the mitochondrion). In *Candida tropicalis*, the cytosolic enzyme provides the primary source of acetoacetyl-CoA for sterol biosynthesis [28]. In *S. cerevisiae*, the reaction step is subject to regulation by the intracellular levels of sterols, by transcriptional regulation mediated by late intermediate(s) or product(s)

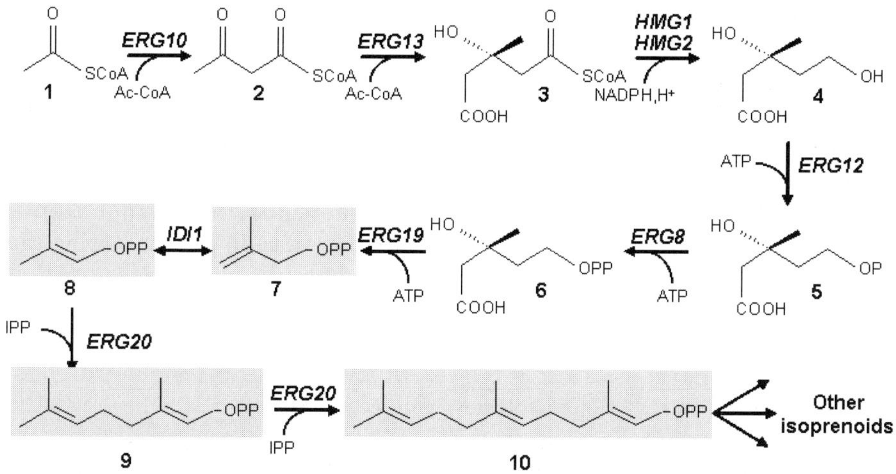

Fig. 3 The mevalonate pathway of *S. cerevisiae* **1**: acetyl-CoA, **2**: acetoacetyl-CoA, **3**: 3-hydroxy-3-methylglutaryl-CoA (HMG-CoA), **4**: mevalonate, **5**: phosphomevalonate, **6**: diphosphomevalonate, **7**: IPP, **8**: DMAPP, **9**: GPP, **10**: FPP. Gray boxes specify the general precursors for the different classes of isoprenoids. The enzymes encoded by the different genes are: *ERG10*: acetoacetyl-CoA thiolase, *ERG13*: HMG-CoA synthase, *HMG1*, *HMG2*: HMG-CoA reductases, *ERG12*: mevalonate kinase, *ERG8*: phosphomevalonate kinase, *ERG19*: diphosphomevalonate decarboxylase, *IDI1*: IPP:DMAPP isomerase, *ERG20*: FPP synthase

of the pathway [29–33]. However, overexpression of *ERG10* did not increase the radiolabeled acetate incorporation on total sterol, suggesting that another enzyme of the sterol biosynthetic pathway is flux-controlling [31].

The condensation of acetyl-CoA with acetoacetyl-CoA to yield 3-hydroxy-3-methylglutaryl-CoA (HMG-CoA) is catalyzed by the *ERG13* gene product, HMG-CoA synthase. This enzymatic step is subject to regulatory control [29, 30]. The details of the regulatory mechanism involved remain uncharacterized [25]. However, the first crystal structure of an HMG-CoA synthase from an organism, *Staphylococcus aureus*, was recently described [34]. Although the staphylococcal and streptococcal enzymes exhibit little similarity (20%) with their eukaryotic counterparts, the amino acid residues involved in the acetylation and condensation reactions are conserved among bacterial and eukaryotic HMG-CoA synthases [34]. The structure provides the molecular basis for a potential reaction mechanism consisting of three steps occurring via a ping-pong mechanism, and provides insight into the rational design of alternative drugs for cholesterol-lowering therapies or novel antibiotic targets for Gram-positive cocci [34].

The third enzyme in the pathway, HMG-CoA reductase, responsible for the conversion of HMG-CoA into mevalonate, is the most studied step of the mevalonate pathway. Unlike humans, *S. cerevisiae* has two copies of the gene encoding HMG-CoA reductase: *HMG1* and *HMG2*, but Hmg1p was shown to be responsible for more than 83% of the enzyme activity in wild type cells [35]. Disruption of both genes renders the cell non-viable, as predicted. This enzymatic step is highly regulated at different levels and appears to be a key regulatory point in the mevalonate pathway.

Mevalonate kinase, encoded by *ERG12*, phosphorylates mevalonate at the C-5 position using ATP. It has been shown that FPP and geranyl diphosphate (GPP) exert an inhibitory effect on the enzyme [36]. The next step catalyzed by the phosphomevalonate kinase, the gene product of *ERG8*, is not subject to feedback regulation by ergosterol [25]. Overexpression of *ERG8* using the strong *GAL1* promoter led to largely unchanged ergosterol levels, suggesting that this enzyme is not flux-controlling for ergosterol production [27].

The next step in the mevalonate pathway involve the *ERG19* gene product (mevalonate diphosphate decarboxylase), which converts mevalonate diphosphate into IPP. The *IDI1* gene product (isopentenyl diphosphate:dimethylallyl diphosphate isomerase) can then convert IPP into dimethylallyl diphosphate (DMAPP). IPP isomerase catalyzes an essential activation step in isoprenoid metabolism in the conversion of IPP to DMAPP by enhancing the electrophilicity of the isoprene unit by at least a billion-fold [37]. Two different classes of IPP isomerases have been reported: the type I enzyme, first characterized in the late 1950s, is widely distributed in eukaryota and eubacteria, while the type II enzyme was recently discovered in *Streptomyces* sp. strain CL190 and in the archaeon *Methanothermobacter thermautotrophicus* [38, 39]. The type I and type II isomerases have different structures and different cofactor requirements, suggesting that they catalyze isomerizations by different chemical mechanisms [38]. The properties of mevalonate diphosphate decarboxylase and of IPP isomerase are largely uncharacterized. However, reduced sterol content observed after overexpression of *ERG19* was attributed to the accumulation of diphosphate intermediates leading to feedback inhibitions [40]. Hence, *ERG19* could encode a flux-controlling step of the mevalonate pathway [40].

The final step in the early portion of the pathway is the conversion of DMAPP into geranyl and farnesyl diphosphates (GPP and FPP, respectively). Farnesyl (geranyl) diphosphate synthase, the product of the *ERG20* gene, catalyzes this reaction. The enzyme first combines DMAPP and IPP to form GPP, and then GPP is extended by combination with a second IPP to form FPP. FPP synthase is a well characterized prenyltransferase. The enzyme has been purified to homogeneity from several eukaryotic sources including *S. cerevisiae* [41], avian liver [42], porcine liver [43, 44] or human liver [45]. FPP is a pivotal molecule situated at the branch point of several important metabolic pathways leading to sterol, heme, dolichol or quinone biosynthesis and prenylation of proteins, and is also involved in several key regulations of the mevalonate pathway. Furthermore, overexpression of *ERG20* has been shown to result in increased levels of enzyme activity and ergosterol production, indicating that FPP synthase may be a flux controlling enzyme [25].

The principal properties of the enzymes of the mevalonate pathway are summarized in Table 4.

The regulation of the isoprenoid biosynthetic pathway is known to be complex in all eukaryotic organisms examined, including the budding yeast *S. cerevisiae* [73–75]. The overriding principle for the regulation of this pathway is multiple levels of feedback inhibition (Fig. 4). This feedback regulation involves several intermediates and appears to act both at different steps of the pathway and at different levels of regulation, as it involves changes in gene transcription, mRNA translation, enzyme activity and protein stability. The emerging picture is that the isoprenoid pathway has a number of points of regulation that act to control the overall flux through the pathway as well as the relative flux through the various branches of the pathway [33]. From these complex multilevel regulations, two distinct but interconnected major sites of regulation are evident: one is the HMG-CoA reductase, the other is due to enzymes competing for FPP.

The yeast HMG-CoA reductase is subject to complex regulation by a number of factors and conditions, at different levels. At the transcriptional level, *HMG1* expression is stimulated by heme via the transcriptional regulator Hap1p, while *HMG2* expression is inhibited, indicating a relationship between heme and sterol biosynthesis [76]. Dimster-Denk et al. [77], showed that Hmg1p was translationally repressed by a non-sterol product of the pathway. In a different study, the same group reported the induction of *HMG1*

Table 4 Properties of the enzymes of the mevalonate pathway of *S. cerevisiae*

Gene	Enzyme	E.C. number	Catalytic properties S.A.	K_m	Cofactors	Metals	Crystal structure	Ref.
ERG10	Acetoacetyl-CoA thiolase	2.3.1.9	$59.8^{\dagger}$ $29^{\dagger}$	$0.77^{a\dagger}$ $1.05^{a\dagger}$		$Ca^{2+\dagger\dagger}$ $Mg^{2+\dagger\dagger}$	[46–48] †††	[49–51]
ERG13	HMG-CoA synthase	2.3.3.10	2.1 2	0.01^{a} 0.0001^{b} 0.01^{a} 0.003^{b}			[34, 52]‡	[53, 55]
HMG1, *HMG2*	HMG-CoA reductase	1.1.1.34	0.0035 0.0038* 0.00058**		NADPH		[56]‡‡	[35, 57]
ERG12	Mevalonate kinase	2.7.1.36	0.77	7.4^{c}	ATP	Ca^{2+} Co^{2+} Fe^{2+} Mg^{2+} Zn^{2+}	[58]‡‡‡	[59–61]

S.A.: Specific activity expressed as $\mu mol\,min^{-1}mg^{-1}$, K_m expressed as mM. †: *Candida tropicalis*, ††: *Rhizobium* sp., †††: *Zooglea ramigera*, ‡: *Staphylococcus aureus*, ‡‡: Human,‡‡‡: *Methanococcus jannaschii*,·: *Streptococcus pneumoniae*, ··: *Escherichia coli*, ···: *Bacillus subtilis*,*: Hmg1p, **: Hmg2p, a: acetyl-CoA, b: acetoacetyl-CoA, c: ATP, d: IPP, e: DMAPP

Table 4 (continued)

Gene	Enzyme	E.C. number	S.A.	Catalytic properties K_m	Cofactors	Metals	Crystal structure	Ref.
ERG8	Phosphomevalonate kinase	2.7.4.2	0.06		ATP	Co^{2+} Fe^{2+} Mg^{2+} Mn^{2+} Zn^{2+}	[62]·	[63]
ERG19	Diphosphomevalonate decarboxylase	4.1.1.33			ATP		[64]	
IDI1	IPP isomerase	5.3.3.2		0.03–0.04[d]			[65, 66]·· [67]···	[68–70]
ERG20	FPP synthase	2.5.1.10	5.22	0.008[e]				[41, 71]
			2.33	0.004-0.01[d]				[72]

S.A.: Specific activity expressed as $\mu mol\,min^{-1} mg^{-1}$, K_m expressed as mM. †: *Candida tropicalis*, ††: *Rhizobium* sp., †††: *Zooglea ramigera*, ‡: *Staphylococcus aureus*, ‡‡: Human, ‡‡‡: *Methanococcus jannaschii*, · : *Streptococcus pneumoniae*, ··: *Escherichia coli*, ···: *Bacillus subtilis*, *: Hmg1p, **: Hmg2p, [a]: acetyl-CoA, [b]: acetoacetyl-CoA, [c]: ATP, [d]: IPP, [e]: DMAPP reporter gene after inhibition of squalene synthase or lanosterol demethylase, suggesting that *HMG1* responded to the levels of sterol products of the pathway [33]. The two yeast isozymes also have distinctly different post-translational fates: Hmglp was shown to be extremely stable while Hmg2p was subject to rapidly regulated degradation depending on the flux through the mevalonate pathway [78]. The stability of each isozyme is determined by its non-catalytic amino-terminal domain. Hmg2p was demonstrated to undergo ERAD (endoplasmic reticulum-associated degradation), similar to its mammalian ortholog, dependent on ubiquitination [78–81]. FPP was demonstrated as the source of the regulatory signal controlling and coupling ubiquitination/degradation of Hmg2p with the flux in the mevalonate pathway [78, 81, 82]. In addition to the FPP signal, an oxysterol-derived signal positively regulates Hmg2p degradation in yeast, but in contrast with mammals it is not an absolute requirement for degradation in yeast [83]. In a recent article, Shearer et al. [80] detailed the basis of ERAD towards Hmg2p.

To summarize, the different regulations of HMG-CoA reductase can be grouped as 1) feedback inhibition (regulation of HMG-CoA reductase activity in response to intermediates or products from the mevalonate pathway), and 2) cross-regulation (regulation by processes independent of the mevalonate pathway) [74]. As a consequence, in aerobic conditions Hmg1p is actively synthesized and extremely stable consistent with the constant need for sterols, while in anaerobic conditions the enzyme with a high turnover, Hmg2p, is dominant in order to allow rapid adjustment of the balance between cellular demand and the potential accumulation of toxic compounds [74]. *HMG1* and *HMG2* are also expressed differently as a function of the growth phase [76, 84].

FPP, the product of FPP synthase (Erg20p), is a pivotal intermediate in the mevalonate pathway leading to the synthesis of several critical end products [25]. In addition, the farnesyl units and the related geranyl and geranylgeranyl species are important elements for the posttranslational modification of proteins that require hydrophobic membrane anchors for proper placement and function. Furthermore, farnesol (FOH), a metabolite causing apoptotic cell death in human acute leukemia, a molecule involved in quorum sensing in *Candida albicans* [85, 86] and causing growth inhibition in *S. cerevisiae*, is endogenously generated in the cells by enzymatic dephosphorylation of FPP [87–89]. To ensure constant production of the multiple isoprenoid compounds at all stages of growth whilst preventing accumulation of potentially toxic intermediates, cells must precisely regulate the level of activity of enzymes of the mevalonate pathway [90]. A number of experimental data show that biosynthesis of dolichols and ubiquinones, as well as isoprenylated proteins, is regulated by enzymes distal to HMG-CoA reductase [91, 92]. This is illustrated on one hand by recent data on the effects of modulating FPP pools on dolichol biosynthesis and on the other hand by effects of increased tRNA prenylation on FPP synthase levels.

In aerobic conditions, a strain with *ERG20* on a multicopy plasmid was characterized by almost six-fold higher FPP synthase activity than a control wild-type strain. Simultaneously, the HMG-CoA reductase activity was changed by about 20%, which is consistent with the known regulations of HMG-CoA reductase activity [91]. Such an immense increase in FPP synthase activity correlated with a significant elevation in dolichol and ergosterol synthesis (about 80% and 32% higher, respectively). These results suggested that FPP synthase, independently of HMG-CoA reductase, is responsible for the partition of FPP, the substrate for squalene synthase and *cis*-prenyltransferase, between the syntheses of both groups of compounds acting as a flux-controlling enzyme [91]. An intricate correlation between FPP synthase activity, ergosterol level and physiology of the cells has also been observed [93]. Nevertheless, the disruption of the squalene synthase gene (when the strain deleted of *ERG9* was cultivated in the presence of ergosterol) resulted in concurrently diminished activities of both FPP synthase and HMG-CoA reductase (78 and 83% repression, respectively). This strongly indicated the implication of squalene synthase in determining the intermediate flow rates in the mevalonate pathway; in other words, when the early intermediates of the pathway cannot be converted to ergosterol and its esters, and synthesis of dolichols is unable to assimilate the bulk of FPP, both FPP synthase and HMG-CoA reductase are repressed [91]. Moreover, changing a *erg9* deleted strain from a medium containing to a medium deprived of ergosterol resulted in a more than ten-fold increase in FPP synthase activity, while HMG-CoA reductase activity was increased by 1.4-fold. Therefore, evidence is given that earlier literature data indicating strictly coordinated regulation of the mevalonate pathway enzymes, i.e. HMG-CoA reductase, FPP synthase, and squalene synthase with HMG-CoA reductase as the main regulatory enzyme in sterol biosynthesis, does not find full confirmation. FPP synthase, independently of HMG-CoA reductase and to a certain degree of squalene synthase, responds the most to changes in internal and external environmental conditions [91]. This is perhaps not surprising if one considers the diversified cell functions in which its product, FPP, directly participates [91].

DMAPP, the substrate of FPP synthase, forms a branch point of the isoprenoid pathway because it is also a substrate of Mod5p, tRNA isopentenyltransferase [94]. As a consequence, tRNA and the isoprenoid biosynthetic pathway compete for DMAPP as a common substrate. It has been shown that overexpression of *ERG20* causes a decrease of i^6A modification of tRNA, so tRNA processing is dependent upon changes in the level of FPP synthase [95]. Moreover, in a strain defective in Maf1p (a negative regulator of tRNA transcription), an excessive amount of DMAPP is dedicated to tRNA modification and, consequently, a lower amount of DMAPP is acces-

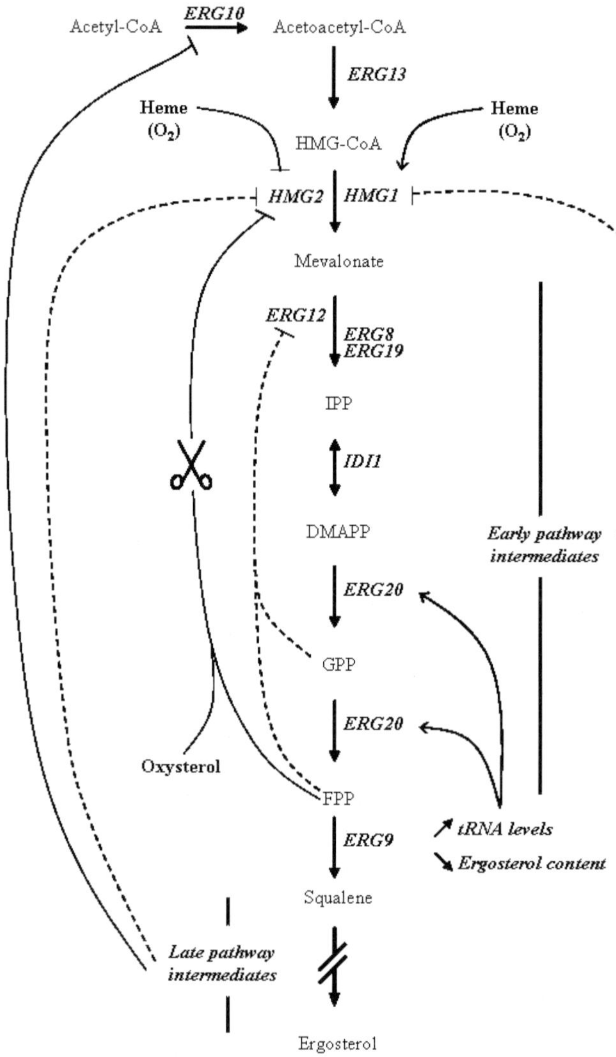

Fig. 4 Principal regulations of the mevalonate pathway. *Straight lines*: regulations at gene expression level, *dashed lines*: regulations at protein synthesis level, ✂: regulation of protein stability

sible for FPP synthase. As a consequence, the *maf1-1* strain is characterized by elevated levels of Erg20p and decreased ergosterol content. In this case, regulation of Erg20p levels is due to both transcriptional and post-translational regulations [95]. Therefore, in yeast, tRNA levels appear to contribute to the complex regulation of FPP synthase and that of the mevalonate pathway.

2.3
The MEP Pathway

Since the discovery of the mevalonate pathway, it has been largely accepted that IPP and DMAPP originated exclusively from this pathway in all living organisms. However, inconsistencies between several results, mainly involving labeling experiments, with the sole operation of the mevalonate pathway have been reported [96–99]. The existence of a second pathway was discovered relatively recently by the research groups of Rohmer and Arigoni using stable isotope incorporation in various eubacteria and plants [15, 18]. These data suggested that pyruvate and a triose phosphate could serve as precursors for the formation of IPP and DMAPP [15]. The gene encoding the first reac-

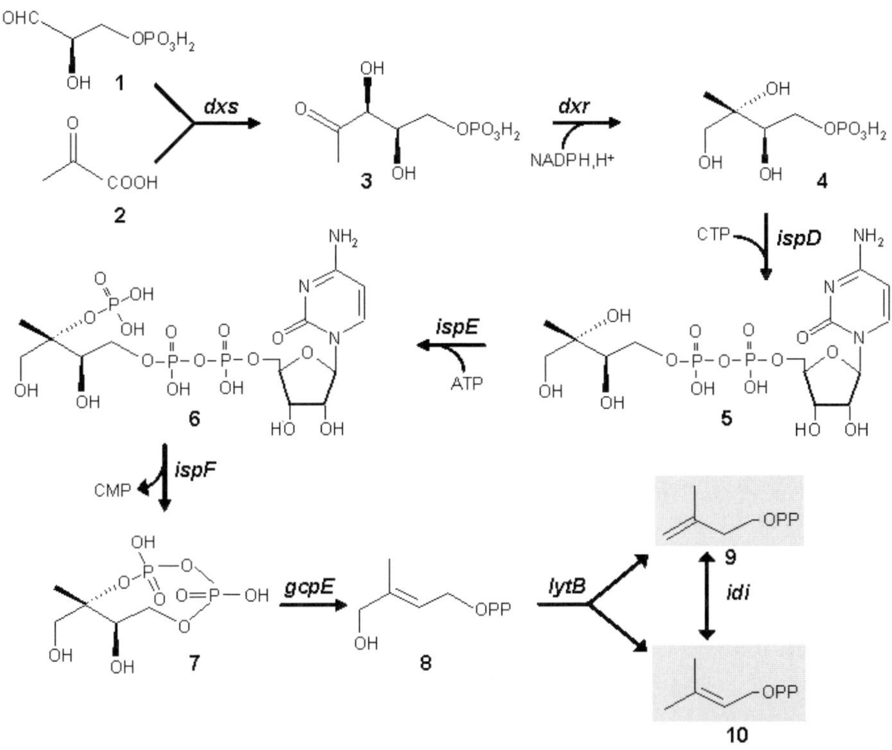

Fig. 5 The *E. coli* MEP pathway for the synthesis of IPP and DMAPP **1**: D-glyceraldehyde 3-phosphate, **2**: pyruvate, **3**: 1-deoxy-D-xylulose 5-phosphate, **4**: 2-C-methyl-D-erythritol 4-phosphate, **5**: 4-diphosphocytidyl-2-C-methyl-D-erythritol, **6**: 2-phospho-4-diphosphocytidyl-2-C-methyl-D-erythritol, **7**: 2-C-methyl-D-erythritol 2,4-cyclodiphosphate, **8**: 1-hydroxy-2-methyl-2-(*E*)-butenyl 4-diphosphate, **9**: isopentenyl diphosphate, **10**: dimethylallyl diphosphate. The enzymes encoded by the different genes are: *dxs*: DXP synthase, *dxr*: DXP isomeroreductase, *ispD*: MEP cytidylyltransferase, *ispE*: CDP-ME kinase, *ispF*: MECDP synthase, *gcpE*: MECDP reductase, *lytB*: HMBPP reductase

tion step of the alternative non-mevalonate pathway was identified and cloned from *E. coli* and the plant *Mentha piperita* [100–102] (Fig. 5). It now seems apparant that most Gram-negative bacteria and *Bacillus subtilis* use the MEP pathway for isoprenoid biosynthesis, whereas staphylococci, streptococci, enterococci, fungi and archaea use the mevalonate pathway [103–106]. Although most *Streptomyces* strains are equipped with the MEP pathway, some of them have been reported to possess the mevalonate pathway in addition to the MEP pathway used to produce terpenoid antibiotics [107–110]. *Listeria monocytogenes* was reported as the only pathogenic bacterium known to contain both pathways concurrently [111]. Plants use the MEP pathway in plastids and the mevalonate pathway in their cytosol. Elucidation of the MEP pathway has been achieved through multidisciplinary approaches including organic chemistry, microbial genetics, biochemistry, molecular biology, and bioinformatics. The impressively rapid increase in information available about the MEP pathway is a good example of the integration of genomics with more traditional approaches to identifying whole metabolic pathways in distant organisms [112].

In the first step of the MEP pathway, 1-deoxy-D-xylulose 5-phosphate synthase, also named DXP synthase or Dxs, catalyzes the condensation of the two precursors from the central metabolism, D-glyceraldehyde 3-phosphate (GAP) and pyruvate, to form DXP. However, DXP synthase is not the first specific enzymatic step of the MEP pathway as, in addition to IPP and DMAPP, DXP is the precursor for the biosynthesis of vitamins B1 (thiamine) and B6 (pyridoxal) in *E. coli* [100]. DXP synthase activity, which is relatively high compared to the other enzymes of the pathway, requires both thiamine and a divalent cation (Mg^{2+} or Mn^{2+}) [113] (Table 5). DXP synthases represent a new class of thiamine diphosphate dependent enzymes combining the characteristics of decarboxylases and transketolases [114].

As DXP is the precursor for different kinds of compounds, the committed step of the pathway is catalyzed by DXP isomeroreductase (Dxr) and leads to the formation of 2-C-methyl-D-erythritol 4-phosphate (MEP), hence its name: "MEP pathway". Takahashi et al. [115] cloned the gene *yaeM* from *E. coli*, and showed that it was responsible for the rearrangement and reduction of DXP in a single step. The gene *yaeM* was therefore renamed *dxr*. The catalytic activity of DXP isomeroreductase is substantially lower ($12\ \mu mol\ mg^{-1} min^{-1}$) than DXP synthase [113] (Table 5). Kuzuyama et al. [116], studying various mutants of DXP isomeroreductase, defined Glu^{231}, Gly^{14}, and three histidine residues (His^{153}, His^{209} and His^{257}) as determining residues for the catalysis. The reaction catalyzed by DXP isomeroreductase is reversible although the equilibrium is largely displaced in favor of the formation of MEP [117]. Due to the wide distribution of DXP isomeroreductase in plants and many eubacteria, including pathogenic bacteria, and its absence in mammalian cells, this enzyme has been studied as a target for herbicides and antibacterial drugs. Fosmidomycin, an antibacterial agent ac-

Table 5 Properties of the enzymes of the MEP pathway

Gene	Enzyme	E.C. number	Catalytic properties S.A.	K_m	Cofactors	Metals	Crystal structure	Ref.
dxs	DXP synthase	2.2.1.7	300 370	96^a, 250^b 65^a, 120^b	TPP	Mg^{2+}		[101,113, 151]
ispC/dxr	DXP isomeroreductase	1.1.1.267	11.8 19.5	$60–250^c$,$7–20^d$ 115^c, 0.5^d 300^c, 5^d	NADPH	Co^{2+} Mn^{2+} Mg^{2+}	[67,152, 153]	[115,116, 119,152, 154,155]
ispD	MEP cytidylyltransferase	2.7.7.60	20–70	131^e, 3.1^f	CTP	Mg^{2+}, Mn^{2+}, Co^{2+}	[156,157]	[113,121, 122]
ispE	CDP-ME kinase	2.7.1.148	33		ATP	Mg^{2+}	[139]	[124–126, 158]
ispF	MECDP synthase	4.6.1.12				Mg^{2+}, Mn^{2+}	[123,139, 159,160]	[113,127, 128]
ispG/gcpE	MECDP reductase	1.17.4.3	0.6	420		Fe^{2+}		[113,141, 161,162]
ispH/lytB	HMBPP reductase	1.17.1.2	6.6	590	NAD(P)H, FAD	Co^{2+}, Fe^{2+}, Mn^{2+}		[142,144, 161]

S.A.: Specific activity expressed as $\mu mol\,min^{-1}mg^{-1}$, K_m is expressed as µM. [a]: pyruvate, [b]: GAP, [c]: DXP, [d]: NADPH, [e]:2*C*-methyl-D-erythritol 4-phosphate, [f]: CTP tive against most Gram-negative and some Gram-positive bacteria, has been shown to be a strong, specific and competitive inhibitor of DXP isomeroreductase activity [115]. For more data about DXP isomeroreductase, see [118].

In order to study the MEP pathway, *E. coli* strains were engineered to allow the study of mutations in otherwise essential genes. For this purpose, in addition to the MEP pathway, *E. coli* was transformed with the genes encoding mevalonate kinase, phosphomevalonate kinase and diphosphomevalonate decarboxylase. This allowed the study of mutants of the MEP pathway which would have led to the lethality of wild-type cells [119, 120]. Mutants with a defect in the synthesis of IPP from MEP were isolated and the genes responsible for this defect identified. These genes are *ygbP*, *ychB*, *ygbB* and *gcpE*. The genes *ygbP*, *ychB*, and *ygbB* are all essential in *E. coli* and the enzymatic steps catalyzed by their gene products belong to the trunk line of the MEP pathway [120].

ygbP (*ispD*) was shown to encode MEP cytidylyltransferase converting MEP into 4-diphosphocytidyl-2-C-methyl-D-erythritol (CDP-ME) in the presence of CTP [121, 122]. Its activity is also substantially lower than DXP synthase activity (Table 5). The dominant feature of its active site is the preponderance of basic side chains involved in binding and processing substrates; in particular, four basic residues were shown to be major contributors for the enzyme mechanism and are strictly conserved: Arg^{20}, Lys^{27}, Arg^{157} and Lys^{213} [123].

In the presence of ATP, CDP-ME is converted into 2-phospho-4-diphosphocytidyl-2-C-methyl-D-erythritol (CDP-ME2P) by the CDP-ME kinase encoded by *ispE* [124, 125]. On the basis of sequence comparisons, CDP-ME kinase was recognized as a member of the GHMP kinase family, which initially included galactose kinase, homoserine kinase, mevalonate kinase and phosphomevalonate kinase, as well as more recently mevalonate 5-diphosphate decarboxylase and the archaeal shikimate kinase [126].

2-C-methyl-D-erythritol 2,4-cyclodiphosphate (MECDP) synthase, encoded by *ygbB* (*ispF*), was demonstrated to catalyze the formation of MECDP from CDP-ME2P with concomitant elimination of cytidine-monophosphate (CMP) [127, 128]. *ispF* has been shown to be essential [120, 129] and conditional mutation of *ispF* in *E. coli* or of its ortholog *yacN* in *B. subtilis* led to a decrease in growth rate and altered cell morphology [130]. In contrast to the dispersed nature of genes belonging to the MEP pathway, *ispD* and *ispF* are transcriptionally coupled or, in some cases, fused into one coding region leading to a bifunctional enzyme. IspDF coupling is highly unusual, as these enzymes catalyze nonconsecutive steps of the MEP pathway. Interactions have been observed between the bifunctional IspDF and IspE protein. Monofunctional IspD, IspF and IspE proteins have also demonstrated a close interaction, suggesting a multienzymatic complex possibly responsible for metabolic flux control through the MEP pathway [131].

In contrast to the mevalonate pathway, in which DMAPP is synthesized from IPP by the essential IPP:DMAPP isomerase activity, the finding that IPP:DMAPP isomerase was functional but non-essential for growth of *E. coli* indicated that the MEP pathway was branched, so that DMAPP and IPP are synthesized by two different routes, splitting at late stages of the pathway [132]. The first evidence for the possible branching of the pathway came from the finding of differential deuterium retention of isoprene units derived from either DMAPP or IPP [133, 134].

The last two steps of the pathway were recently solved by Hintz et al. [135], who reported the accumulation of the formerly unknown intermediate 1-hydroxy-2-methyl-2-(*E*)-butenyl 4-diphosphate (HMBPP) in a *lytB* (*ispH*) disrupted *E. coli* strain. Several studies aimed at demonstrating the essential nature of *gcpE* (*ispG*) and/or *lytB* [136, 137], their necessity for DXP conversion to IPP and DMAPP [138–140], and the efficiency of their gene products in converting MECDP into HMBPP [141] and HMBPP into IPP and DMAPP [142]. An important feature of both GcpE and LytB is a [4Fe – 4S] cluster present as a prosthetic group, underlying their high sensitivity towards oxygen. This property, common to both enzymes, may explain why investigations of the terminal reactions of the MEP pathway have been hampered for so long [141, 143, 144]. No X-ray crystal structure is available for GcpE; however, Brandt et al. [145] developed a model for part of GcpE from *Streptomyces coelicolor*, reported to contain the active site. Although the natural cofactors and electron donors of GcpE and LytB remain to be elucidated, the main steps of the MEP pathway appear to have been clearly demonstrated.

The finding that a single enzyme is responsible for the formation of both IPP and DMAPP contrasts with the mevalonate pathway where DMAPP is successively formed from IPP by IPP isomerase. As a consequence of these findings, the role of IPP isomerase in microorganisms expressing the MEP pathway comes into question. The non-essential and non-limiting roles of IPP isomerase activity are currently being investigated, as on the one hand, the *E. coli* Idi enzyme was reported to have 20-fold less activity than its yeast counterpart [146], *idi* from *E. coli* is dispensable [132] and *idi* homologs have not been found in genomes of many bacteria using the MEP pathway sequenced so far [147]; on the other hand, structurally and mechanistically different IPP isomerases, referred to as class II IPP isomerases, have been identified in *Streptomyces* sp. strain CL190 and also in a variety of Gram-positive bacteria, cyanobacteria and archaebacteria [108]. Furthermore, the overexpression of *idi* genes of different origins in *E. coli* engineered for the production of lycopene has always led to carotenoid overproduction [147–149]; these findings fuel the debate about the non-essentiality and non-limiting role of the IDI reaction [150].

3
Metabolic Engineering of Microorganisms for Isoprenoid Production

In the last decade there have been a number of investigations into the construction of engineered microorganisms with the ability to produce different isoprenoids. Fig. 6 schematically shows the different steps for constructing industrial isoprenoid-producing microorganisms. As we will see in the next sections, a common feature for most of the studies conducted on microbial isoprenoid production is that they include expression of heterologous genes for converting isoprenoid precursors of the host microorganism into the desired isoprenoid, and deregulation of metabolic pathways in order to increase the metabolic flux to isoprenoid precursors.

Tetraterpenoid carotenoids (C_{40}) have been the most interesting group of isoprenoids for metabolic engineering because of their easy color screening [163] and their industrial importance as feed supplements in the poultry and fish farming industries [164]. The carotenoid biosynthetic pathway in *Erwinia uredovora* was first elucidated by Misawa et al. [165], and the corresponding genes were subsequently used in several studies for production of heterologous carotenoids in non-carotenogenic microorganisms. However, isolation and characterization of more than 150 carotenogenic genes involved in the synthesis of 27 different enzymes in the carotenoid biosynthesis pathways in different organisms [166, 167] has opened the door to the heterologous production of a broad range of carotenoids.

Ergosterol (the main sterol in yeasts), found in large amounts in yeast membranes, plays a key role in regulating the membrane fluidity and permeability [168], and is produced through the mevalonate pathway. Although *E. coli* has been the main host for metabolic engineering of isoprenoids, in

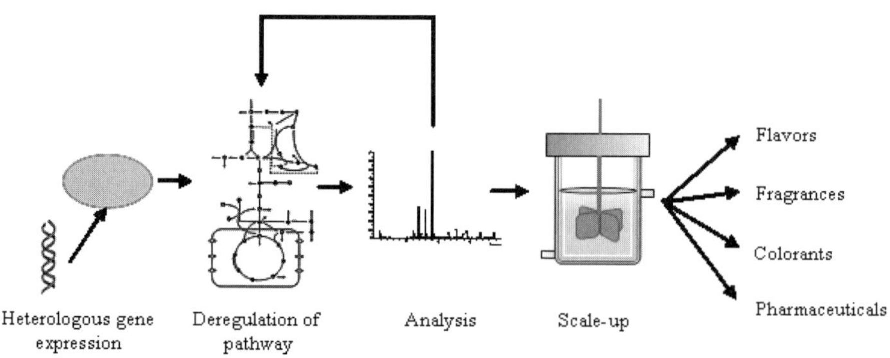

Fig. 6 Summary of different steps for establishing industrial cell factories capable of isoprenoid production

some cases yeasts (which have high capacity for ergosterol production) have been subject to metabolic engineering studies [169–172].

3.1
Metabolic Engineering of the MEP Pathway

Amongst the different enzymes in the MEP pathway, DXP synthase (encoded by *dxs*), IPP isomerase (encoded by *idi*) and DXP isomeroreductase (encoded by *dxr*) have been the main targets for metabolic engineering investigations. Overexpression of *dxs* has been achieved in several studies in order to improve the intracellular pool of precursors for isoprenoid biosynthesis [173–181]. For example, overexpression of *dxs* in *E. coli* strains harboring the carotenogenic genes resulted in up to 10.8- and 3.9-fold increases in the accumulated levels of lycopene and zeaxanthin, respectively [178]. Overproduction of DXP synthase also had a great impact on the biosynthesis of taxadiene [173] as the required intermediate for the synthesis of paclitaxel (Taxol), known as the most important anti-cancer drug introduced in the last ten years [182]. Harker & Bramley [179] also showed elevated levels of lycopene in engineered *E. coli* upon overexpression of *dxs*. Kim & Keasling [180] noticed the importance of promoter strength and plasmid copy number in balancing expression of *dxs* with overall metabolism.

The second step in the MEP pathway, which is catalyzed by DXP isomeroreductase, has been shown to control the flux to isoprenoid precursors in *E. coli* [180, 181]. Co-overexpression of *dxr* and *dxs* was concomitant with a 1.4- to 2-fold increase in lycopene level compared to the strains overexpressing only *dxs* [180]. However, overexpression of *dxs* had a greater impact on lycopene production than overexpression of *dxr*. In another study [181], simultaneous overexpression of *dxs* and *dxr* in the β-carotene- and zeaxanthin-producing *E. coli* strains was lethal for the cells, probably due to restricted storage capacity for lipophilic carotenoids, which causes membrane overload and loss of functionality. This problem implies the need for host microorganisms with higher storage capacity for heterologous production of carotenoids [24, 183, 184].

Isomerization of IPP to DMAPP has been another target for improving isoprenoid biosynthesis in the MEP pathway, and several studies have shown the enhancing effect of IPP isomerase overproduction [148, 149, 173, 174, 176, 181]. Overexpression of *idi* genes from different organisms in recombinant *E. coli* showed 1.5- to 4.5-fold increases in the lycopene, β-carotene, and phytoene levels compared to the control strains [148]. Positive effects of *idi* or *dxs* overexpression on β-carotene and zeaxanthin accumulation in *E. coli* have also been shown. Amplification of *idi* or/and *dxs* gave approximately 2–3 times more carotenoid accumulation in the recombinant strains than the control [181]. Engineered lycopene-producing *E. coli* overexpressing *dxs, idi,* and *ispA* (responsible for FPP synthase activity in *E. coli*) produced six-fold more lycopene than the control strain [174]. Simultaneous amplification of *idi* and GGPP synthase gene (*gps*) in astaxanthin-producing *E. coli* strains increased the astaxanthin level from 33 μg/g dry weight in the control strain to 1419 μg/g dry weight in the recombinant strain [149]. In the same laboratory, subjecting the *gps* gene to direct evolution resulted in a two-fold increase in the lycopene level, and subsequent cooverexpression of the *dxs* gene further enhanced the lycopene accumulation [177].

The MEP pathway is initiated with the combination of pyruvate and GAP in equal amounts, catalyzed by DXP synthase. Hence, balanced pools of pyruvate and GAP would be an important factor in the efficient direction of the central carbon metabolism to the isoprenoid pathway. Pyruvate is required as a precursor in many cellular pathways and it is presumably more available than GAP for isoprenoid biosynthesis. It was shown that overproduction or inactivation of enzymes that leads to redirection of flux from pyruvate to GAP results in higher lycopene production in *E. coli* [185]. Thus, overproduction of phosphophenolpyruvate (PEP) synthase (Pps) and PEP carboxykinase (Pck) or inactivation of pyruvate kinase isozymes (Pyk-I and Pyk-II) were shown to enhance lycopene production in *E. coli*.

Poor expression of plant genes and inadequate amounts of enzymes could be another limiting factor for the production of plant isoprenoids in the engineered hosts [175]. To circumvent the problems of low sesquiterpene yield that arise from the poor expression of plant genes, in one study [176], a codon-optimized variant of amorphadiene synthase gene (*ADS*) was synthesized and expressed in *E. coli*. This improved the enzyme synthesis and production yield of amorphadiene and changed the flux control in the biosynthesis of sesquiterpenes from the step catalyzed by the heterologous plant genes to the supply of precursor (FPP) provided by the MEP pathway. The expression of this synthetic *ADS* gene in *E. coli* resulted in a 10- to 300-fold increase in sesquiterpene accumulation compared to the previous study [175] in which the native plant sesquiterpene synthase genes were expressed. Further overexpression of genes responsible for the synthesis of DXP synthase, IPP isomerase and FPP synthase, with the synthetic *ADS*, led to a 3.6-fold increase in the concentration of amorphadiene, indicating that the supply of precursor limits the sesquiterpene production. However, considering the fact that overexpression of three flux-controlling enzymes of the pathway only resulted in a 3.6-fold increase in amorphadiene concentration, this approach to increasing the flux to FPP seems to be limited by some other native control mechanisms in *E. coli*. Introduction of the mevalonate pathway from *S. cerevisiae* to *E. coli* has been shown to be an alternative approach to increasing the intracellular concentration of isoprenoid precursors, thereby circumventing the as-yet unidentified regulations of the native MEP pathway and also minimizing the complicated regulatory network of the mevalonate pathway observed in yeast, and this resulted in a further ten-fold increase in the amorphadiene concentration [176].

3.2
Metabolic Engineering of the Mevalonate Pathway

Engineering of the industrially-important yeasts, *S. cerevisiae* and *Candida utilis*, for carotenoid production, by introducing the carotenoid biosynthetic genes of *E. uredovora* has been reported [169–172]. Modification of carotenogenic genes based on the codon usage of the *C. utilis* GAP dehydrogenase gene, increased the phytoene and lycopene contents of the strains 1.5- and 4-fold, respectively, compared to those of the strains carrying unmodified genes [171]. HMG-CoA reductase is believed to be the key enzyme in the mevalonate pathway, and overexpression of both full-length and truncated versions of the genes responsible for HMG-CoA reductase synthesis increased the lycopene production in *C. utilis*, but the truncated version had greater impact. Subsequent disruption of the *ERG9* gene also improved lycopene production [172]. The stimulating effect of HMG-CoA reductase overproduction on the lycopene and neurosporaxanthin content in a naturally carotenoid-producing fungus, *Neurospora crassa* [186] and on epicedrol production in *S. cerevisiae* [187] have also been shown.

Table 6 summarizes the examples of metabolically-engineered microorganisms for production of different isoprenoids.

3.3
Metabolic Engineering for Heterologous Production of Novel Isoprenoids

Metabolic engineering can also be applied for heterologous microbial production of novel isoprenoids. In the past few years, production of uncommon and non-commercially-available carotenoids has drawn much attention because of the increasingly scientific documents indicating their potential applications in preventing cancer and cardiovascular diseases as well as their anti-tumor properties [188–191]. However, production of these complex carotenoids by chemical synthesis is impractical, and natural sources contain only trace amounts of these carotenoids. Hence, microbial production is the best choice for their commercial production. Expression or combination of carotenogenic genes from different bacteria in *E. coli* was successfully applied to the production of a number of novel hydroxycarotenoids [192, 193]. In another example [194], *E. coli* transformants were developed by introducing seven carotenoid biosynthetic genes from *E. uredovora* and *A. aurantiacum* for the production of new astaxanthin glucosides. Production of two other uncommon acyclic carotenoids has been achieved in *E. coli* by introducing the *crtC* and *crtD* genes from *Rhodobacter* and *Rubrivivax* [195]. Schmidt-Dannert et al. [196] shuffled phytoene desaturases (encoded by *crtI*) and lycopene cyclases (encoded by *crtY*) from different bacterial species to evolve new enzyme functions and produce a library of carotenoids.

Table 6 Examples of different isoprenoids produced by metabolically-engineered microorganisms

Class	Isoprenoid	Host microorganism	Yield/ concentration	Ref.
Monoterpenoids	Limonene	*E. coli*	~ 5000 μg/L	[197]
	3-Carene	*E. coli*	3 μg/L/OD_{600}	[174]
Diterpenoids	Taxadiene	*E. coli*	1300 μg/L	[173]
	Casbene	*E. coli*	30 μg/L/OD_{600}	[174]
Sesquiterpenoids	(+)-δ-Cadinene	*E. coli*	10.3 μg/L	[175]
	5-*Epi*-aristolochene	*E. coli*	0.24 μg/L	[175]
	Vetispiradiene	*E. coli*	6.4 μg/L	[175]
	Amorphadiene	*E. coli*	24 000 μg/L[a]	[176]
	Epi-cedrol	*S. cerevisiae*	370 μg/L	[187]
Carotenoids	Lycopene	*E. coli*	25 000 μg/gDW	[185]
	Lycopene	*E. coli*	1333 μg/gDW	[178]
	Lycopene	*E. coli*	~ 1000 μg/gDW	[179]
	Lycopene	*E. coli*	22 000 μg/L	[180]
	Lycopene	*E. coli*	45 000 μg/gDW	[177]
	Lycopene	*E. coli*	1029 μg/gDW	[148]
	Lycopene	*E. coli*	1210 μg/L	[174]
	Lycopene	*S. cerevisiae*	113 μg/gDW	[169]
	Lycopene	*C. utilis*	758 μg/gDW	[170]
	Lycopene	*C. utilis*	1100 μg/gDW	[171]
	Lycopene	*C. utilis*	7800 μg/gDW	[172]
	Lycopene	*N. crassa*	17.9 μg/gDW	[186]
	β -Carotene	*E. coli*	1310 μg/gDW	[148]
	β -Carotene	*E. coli*	1533 μg/gDW	[181]
	β -Carotene	*S. cerevisiae*	103 μg/gDW	[169]
	β -Carotene	*C. utilis*	400 μg/gDW	[171]
	β -Carotene	*Z. mobilis*	220 μg/gDW	[198]
	β -Carotene	*A. tumefaciens*	350 μg/gDW	[198]
	Astaxanthin	*E. coli*	1419 μg/gDW	[149]
	Astaxanthin	*C. utilis*	400 μg/gDW	[171]
	Zeaxanthin	*E. coli*	289 μg/gDW	[184]
	Zeaxanthin	*E. coli*	592 μg/gDW	[178]
	Zeaxanthin	*E. coli*	1570 μg/gDW	[181]
	Neurosporaxanthin	*N. crassa*	63.4 μg/gDW	[186]

[a] 112 200 μg/L expected if evaporation is taken into account

4 Outlook

This paper charts the attempts made to move towards green chemistry by reviewing recent investigations into isoprenoid production using metabolically-engineered cell factories. Metabolic engineering represents a pivotal toolset for developing green chemistry solutions for the production of various chemicals. However, we are still far from the extensive use of microbial cell factories for the commercial production of isoprenoids. There is a lack of information about the enzymes involved in the biosynthesis of isoprenoids and the mechanisms underlying the immense complex regulatory network of pathways have not been completely elucidated. Despite the crucial importance of metabolic flux analysis (MFA) and metabolic control analysis (MCA) as helpful tools in designing metabolic engineering strategies, there is no reported work on the application of these tools for microbial isoprenoid production. To perform MFA, metabolic fluxes should be measured, and therefore precise and robust analytical techniques will be needed in order to analyze the intracellular metabolites of pathways. Genome-scale metabolic models for the most common microbial hosts in isoprenoid production, *E. coli* [199, 200] and *S. cerevisiae* [201], have been developed in recent years and can be used in the directed manipulation of the cellular network to predict the changes that are required in the genotype of microorganism in order to obtain efficient microbial strains [202].

However, the improvement of microbial strains for isoprenoid production is only one example that shows how metabolic engineering can be applied when developing green chemistry solutions. There is also a great trend towards the engineering of microbial hosts for the commercial production of other metabolites like polyketides, organic acids, amino acids, and so on. It is expected that all aspects of sustainable development—environment, economics and society—will benefit by the development of green chemistry [7]. Reducing dependency on fossil fuels, saving energy, reducing CO_2 emissions, broadening the range of substrates, reducing costs and improving productivity are some of the environmental and economical advantages. Creation of jobs and the development of new technology platforms that address future challenges are the positive impacts on society [7]. New companies are forming that make use of these new technologies. Poalis (www.poalis.dk), Metabolic Explorer (www.metabolic-explorer.com), Fluxome Science (www.fluxome.com), Institute for OneWorld Health (www.oneworldhealth.org), Amyris Biotechnologies (www.amyrisbiotech.com) and Combinature Biopharm AG (www.combinature.com) are a few examples of small start-up companies that have white biotechnologies as their foci and the development of novel bioprocesses as components of their business plans.

References

1. Nielsen J (2001) Appl Microbiol Biot 55:263
2. Ostergaard S, Olsson L, Nielsen J (2001) Biotechnol Bioeng 73:412

3. Thykaer J, Nielsen J (2003) Metab Eng 5:56
4. Stephanopoulos G, Gill RT (2001) Adv Biochem Eng Biotechnol 73:1
5. Burkart MD (2003) Org Biomol Chem 1:1
6. Grabley S, Thiericke R (1999) Adv Biochem Eng Biotechnol 64:101
7. EuropaBio (2003) White biotechnology: Gateway to a more sustainable future. EuropaBio, Lyon. Available at http://www.mckinsey.com/clientservice/chemicals/pdf/BioVision_Booklet_final.pdf
8. Robin J, Jakobsen M, Beyer M, Noorman H, Nielsen J (2001) Appl Microbiol Biotechnol 57:357
9. Sacchettini JC, Poulter CD (1997) Science 277:1788
10. McCaskill D, Croteau R (1997) Adv Biochem Eng Biotechnol 55:107
11. Wallach O (1887) Justus Liebigs Ann Chem 239:1
12. Ruzicka L (1953) Experientia 9:357
13. Katsuki H, Bloch K (1967) J Biol Chem 242:222
14. Lynen F (1967) Pure Appl Chem 14:137
15. Rohmer M, Knani M, Simonin P, Sutter B, Sahm H (1993) Biochem J 295:517
16. Rohmer M (1999) Nat Prod Rep 16:565
17. Broers STJ (1994) PhD thesis, Eidgenössische Technische Hochschule Zürich
18. Schwarz MK (1994) PhD thesis, Eidgenössische Technische Hochschule Zürich
19. Bach TJ, Boronat A, Campos N, Ferrer A, Vollack K-U (1999) Crit Rev Biochem Mol Biol 34:107
20. Koepp AE, Hezari M, Zajicek J, Vogel BS, LaFever RE, Lewis NG, Croteau R (1995) J Biol Chem 270:8686
21. Mukaiyama T, Shiina I, Iwadare H, Saitoh M, Nishimura T, Ohkawa N, Sakoh H, Nishimura K, Tani Y-I, Hasegawa M, Yamada K, Saitoh K (1999) Chem Eur J 5:121
22. Danishefsky SJ, Masters JJ, Young WB, Link JT, Snyder LB, Magee TV, Jung DK, Isaacs RCA, Bornmann WG, Alaimo CA, Coburn CA, Di Grandi MJ (1996) J Am Chem Soc 118:2843
23. Miyaoka H, Honda D, Mitome H, Yamada Y (2002) Tetrahedron Lett 43:7773
24. Sandmann G, Albrecht M, Schnurr G, Knörzer O, Böger P (1999) Trends Biotechnol 17:233
25. Daum G, Lees ND, Bard M, Dickson R (1998) Yeast 14:1471
26. Veen M, Lang C (2004) Appl Microbiol Biot 63:635
27. Lees ND, Bard M, Kirsch DR (1999) Crit Rev Biochem Mol Biol 34:33
28. Kurihara T, Ueda M, Kamasawa N, Osumi M, Tanaka A (1992) J Biochem (Tokyo) 112:845
29. Trocha PJ, Sprinson DB (1976) Arch Biochem Biophys 174:45
30. Servouse M, Karst F (1986) Biochem J 240:541
31. Dimster-Denk D, Rine J (1996) Mol Cell Biol 16:3981
32. Dixon G, Scanlon D, Cooper S, Broad P (1997) J Steroid Biochem Mol Biol 62:165
33. Dimster-Denk D, Rine J, Phillips J, Scherer S, Cundiff P, DeBord K, Gilliland D, Hickman S, Jarvis A, Tong L, Ashby M (1999) J Lipid Res 40:850
34. Campobasso N, Patel M, Wilding IE, Kallender H, Rosenberg M, Gwynn MN (2004) J Biol Chem 279:44883
35. Basson ME, Thorsness M, Rine J (1986) Proc Natl Acad Sci USA 83:5563
36. Dorsey JK, Porter JW (1968) J Biol Chem 243:4667
37. Anderson MS, Muehlbacher M, Street IP, Proffitt J, Poulter CD (1989) J Biol Chem 264:19169
38. Barkley SJ, Cornish RM, Poulter CD (2004) J Bacteriol 186:1811
39. Kaneda K, Kuzuyama T, Takagi M, Seto H (2001) Proc Natl Acad Sci 98:932

40. Bergès T, Guyonnet D, Karst F (1997) J Bacteriol 179:4664
41. Eberhardt NL, Rilling HC (1975) J Biol Chem 250:863
42. Reed BC, Rilling HC (1975) Biochemistry 14:50
43. Barnard GF, Langton B, Popjak G (1978) Biochem Biophys Res Commun 85:1097
44. Yeh LS, Rilling HC (1977) Arch Biochem Biophys 183:718
45. Barnard GF, Popjak G (1981) Biochim Biophys Acta 661:87
46. Modis Y, Wierenga RK (1999) Structure Fold Des 7:1279
47. Modis Y, Wierenga RK (2000) J Mol Biol 297:1171
48. Kursula P, Ojala J, Lambeir AM, Wierenga RK (2002) Biochemistry 41:15543
49. Kanayama N, Himeda Y, Atomi H, Ueda M, Tanaka A (1997) J Biochem (Tokyo) 122:616
50. Kurihara T, Ueda M, Tanaka A (1989) J Biochem (Tokyo) 106:474
51. Kim SA, Copeland L (1997) Appl Environ Microbiol 63:3432
52. Theisen MJ, Misra I, Saadat D, Campobasso N, Miziorko HM, Harrison DH (2004) Proc Natl Acad Sci USA 101:16442
53. Middleton B (1972) Biochem J 126:35
54. Cabano J, Buesa C, Hegardt FG, Marrero PF (1997) Insect Biochem Mol Biol 27:499
55. Middleton B, Tubbs PK (1975) Methods Enzymol 35:173
56. Istvan ES, Deisenhofer J (2001) Science 292:1160
57. Durr IF, Rudney H (1960) J Biol Chem 235:2572
58. Yang D, Shipman LW, Roessner CA, Scott AI, Sacchettini JC (2002) J Biol Chem 277:9462
59. Gray JC, Kekwick RG (1972) Biochim Biophys Acta 279:290
60. Tchen TT (1958) J Biol Chem 233:1100
61. Porter JW (1985) Methods Enzymol 110:71
62. Romanowski MJ, Bonanno JB, Burley SK (2002) Proteins 47:568
63. Bloch K, Chaykin S, Phillips AH, De Waard A (1959) J Biol Chem 234:2595
64. Bonanno JB, Edo C, Eswar N, Pieper U, Romanowski MJ, Ilyin V, Gerchman SE, Kycia H, Studier FW, Sali A, Burley SK (2001) Proc Natl Acad Sci USA 98:12896
65. Durbecq V, Sainz G, Oudjama Y, Clantin B, Bompard-Gilles C, Tricot C, Caillet J, Stalon V, Droogmans L, Villeret V (2001) EMBO J 20:1530
66. Wouters J, Oudjama Y, Ghosh S, Stalon V, Droogmans L, Oldfield E (2003) J Am Chem Soc 125:3198
67. Steinbacher S, Kaiser J, Eisenreich W, Huber R, Bacher A, Rohdich F (2003) J Biol Chem 278:18401
68. Reardpon JE, Abeles RH (1985) J Am Chem Soc 107:4078
69. Agranoff BW, Eggerer H, Henning U, Lynen F (1960) J Biol Chem 235:326
70. Street IP, Poulter CD (1990) Biochemistry 29:7531
71. Rilling HC (1985) Methods Enzymol 110:145
72. Bartlett DL, King CH, Poulter CD (1985) Methods Enzymol 110:171
73. Goldstein JL, Brown MS (1990) Nature 343:425
74. Hampton R, Dimster-Denk D, Rine J (1996) Trends Biochem Sci 21:140
75. Hampton RY (1998) Curr Opin Lipidol 9:93
76. Thorsness M, Schafer W, D'Ari L, Rine J (1989) Mol Cell Biol 9:5702
77. Dimster-Denk D, Thorsness MK, Rine J (1994) Mol Biol Cell 5:655
78. Hampton RY, Rine J (1994) J Cell Biol 125:299
79. Nakanishi M, Goldstein JL, Brown MS (1988) J Biol Chem 263:8929
80. Shearer AG, Hampton RY (2004) J Biol Chem 279:188
81. Hampton RY, Bhakta H (1997) Proc Natl Acad Sci USA 94:12944
82. Gardner RG, Hampton RY (1999) J Biol Chem 274:31671

83. Gardner RG, Shan H, Matsuda SP, Hampton RY (2001) J Biol Chem 276:8681
84. Casey WM, Keesler GA, Parks LW (1992) J Bacteriol 174:7283
85. Hornby JM, Jensen EC, Lisec AD, Tasto JJ, Jahnke B, Shoemaker R, Dussault P, Nickerson KW (2001) Appl Environ Microbiol 67:2982
86. Grabińska K, Palamarczyk G (2002) FEMS Yeast Res 2:259
87. Haug JS, Goldner CM, Yazlovitskaya EM, Voziyan PA, Melnykovych G (1994) Biochim Biophys Acta 1223:133
88. Melnykovych G, Haug JS, Goldner CM (1992) Biochem Biophys Res Commun 186:543
89. Machida K, Tanaka T, Fujita K, Taniguchi M (1998) J Bacteriol 180:4460
90. Brown MS, Goldstein JL (1980) J Lipid Res 21:505
91. Szkopińska A, Świeżewska E, Karst F (2000) Biochem Biophys Res Commun 267:473
92. Grabowska D, Karst F, Szkopińska A (1998) FEBS Lett 434:406
93. Karst F, Plochocka D, Meyer S, Szkopińska A (2004) Cell Biol Int 28:193
94. Gillman EC, Slusher LB, Martin NC, Hopper AK (1991) Mol Cell Biol 11:2382
95. Kamińska J, Grabińska K, Kwapisz M, Sikora J, Smagowicz WJ, Palamarczyk G, Żołądek T, Boguta M (2002) FEMS Yeast Res 2:31
96. Zhou D, White RH (1991) Biochem J 273:627
97. Cane DE, Rossi T, Pachlatko JP (1979) Tetrahedron Lett 20:3639
98. Cane DE, Rossi T, Tillman AM, Pachlatko JP (1981) J Am Chem Soc 103:1838
99. Flesch G, Rohmer M (1988) Eur J Biochem 175:405
100. Sprenger GA, Schörken U, Wiegert T, Grolle S, de Graaf AA, Taylor SV, Begley TP, Bringer-Meyer S, Sahm H (1997) Proc Natl Acad Sci USA 94:12857
101. Lois L-M, Campos N, Putra SR, Danielsen K, Rohmer M, Boronat A (1998) Proc Natl Acad Sci USA 95:2105
102. Lange BM, Wildung MR, McCaskill D, Croteau R (1998) Proc Natl Acad Sci USA 95:2100
103. Wilding EI, Brown JR, Bryant AP, Chalker AF, Holmes DJ, Ingraham KA, Iordanescu S, So CY, Rosenberg M, Gwynn MN (2000) J Bacteriol 182:4319
104. Hedl M, Sutherlin A, Wilding EI, Mazzulla M, McDevitt D, Lane P, Burgner JW, Lehnbeuter KR, Stauffacher CV, Gwynn MN, Rodwell VW (2002) J Bacteriol 184:2116
105. Bochar DA, Stauffacher CV, Rodwell VW (1999) Mol Genet Metab 66:122
106. Doolittle WF, Logsdon JM (1998) Curr Biol 8:209
107. Takagi M, Kuzuyama T, Takahashi S, Seto H (2000) J Bacteriol 182:4153
108. Hamano Y, Dairi T, Yamamoto M, Kawasaki T, Kaneda K, Kuzuyama T, Itoh N, Seto H (2001) Biosci Biotechnol Biochem 65:1627
109. Hamano Y, Dairi T, Yamamoto M, Kuzuyama T, Itoh N, Seto H (2002) Biosci Biotechnol Biochem 66:808
110. Kawasaki T, Kuzuyama T, Furihata K, Itoh N, Seto H, Dairi T (2003) J Antibiot (Tokyo) 56:957
111. Begley M, Gahan CG, Kollas AK, Hintz M, Hill C, Jomaa H, Eberl M (2004) FEBS Lett 561:99
112. Rodríguez-Concepción M, Boronat A (2002) Plant Physiol 130:1079
113. Eisenreich W, Bacher A, Arigoni D, Rohdich F (2004) Cell Mol Life Sci 61:1401
114. Eubanks LM, Poulter CD (2003) Biochemistry 42:1140
115. Takahashi S, Kuzuyama T, Watanabe H, Seto H (1998) Proc Natl Acad Sci USA 95:9879
116. Kuzuyama T, Takahashi S, Takagi M, Seto H (2000) J Biol Chem 275:19928
117. Hoeffler J-F, Tritsch D, Grosdemange-Billiard C, Rohmer M (2002) Eur J Biochem 269:4446
118. Proteau PJ (2004) Bioorg Chem 32:483

119. Kuzuyama T (2002) Biosci Biotechnol Biochem 66:1619
120. Campos N, Rodríguez-Concepción M, Sauret-Güeto S, Gallego F, Lois L-M, Boronat A (2001) Biochem J 353:59
121. Rohdich F, Wungsintaweekul J, Fellermeier M, Sagner S, Herz S, Kis K, Eisenreich W, Bacher A, Zenk MH (1999) Proc Natl Acad Sci USA 96:11758
122. Kuzuyama T, Takagi M, Kaneda K, Dairi T, Seto H (2000) Tetrahedron Lett 41:703
123. Hunter WN, Bond CS, Gabrielsen M, Kemp LE (2003) Biochem Soc Trans 31:537
124. Lüttgen H, Rohdich F, Herz S, Wungsintaweekul J, Hecht S, Schuhr CA, Fellermeier M, Sagner S, Zenk MH, Bacher A, Eisenreich W (2000) Proc Natl Acad Sci USA 97:1062
125. Kuzuyama T, Takagi M, Kaneda K, Dairi T, Seto H (2000) Tetrahedron Lett 41:2925
126. Miallau L, Alphey MS, Kemp LE, Leonard GA, McSweeney SM, Hecht S, Bacher A, Eisenreich W, Rohdich F, Hunter WN (2003) Proc Natl Acad Sci USA 100:9173
127. Takagi M, Kuzuyama T, Kaneda K, Dairi T, Seto H (2000) Tetrahedron Lett 41:3395
128. Herz S, Wungsintaweekul J, Schuhr CA, Hecht S, Lüttgen H, Sagner S, Fellermeier M, Eisenreich W, Zenk MH, Bacher A, Rohdich F (2000) Proc Natl Acad Sci USA 97:2486
129. Freiberg C, Wieland B, Spaltmann F, Ehlert K, Brotz H, Labischinski H (2001) J Mol Microbiol Biotechnol 3:483
130. Campbell TL, Brown ED (2002) J Bacteriol 184:5609
131. Gabrielsen M, Bond CS, Hallyburton I, Hecht S, Bacher A, Eisenreich W, Rohdich F, Hunter WN (2004) J Biol Chem
132. Rodríguez-Concepción M, Campos N, Maria LL, Maldonado C, Hoeffler J-F, Grosdemange-Billiard C, Rohmer M, Boronat A (2000) FEBS Lett 473:328
133. Giner J-L, Jaun B, Arigoni D (1998) J Chem Soc Chem Commun 1857
134. Charon L, Hoeffler J-F, Pale-Grosdemange C, Lois L-M, Campos N, Boronat A, Rohmer M (2000) Biochem J 346:737
135. Hintz M, Reichenberg A, Altincicek B, Bahr U, Gschwind RM, Kollas AK, Beck E, Wiesner J, Eberl M, Jomaa H (2001) FEBS Lett 509:317
136. Altincicek B, Kollas AK, Sanderbrand S, Wiesner J, Hintz M, Beck E, Jomaa H (2001) J Bacteriol 183:2411
137. Altincicek B, Kollas A, Eberl M, Wiesner J, Sanderbrand S, Hintz M, Beck E, Jomaa H (2001) FEBS Lett 499:37
138. Hecht S, Eisenreich W, Adam P, Amslinger S, Kis K, Bacher A, Arigoni D, Rohdich F (2001) Proc Natl Acad Sci USA 98:14837
139. Steinbacher S, Kaiser J, Wungsintaweekul J, Hecht S, Eisenreich W, Gerhardt S, Bacher A, Rohdich F (2002) J Mol Biol 316:79
140. Campos N, Rodríguez-Concepción M, Seemann M, Rohmer M, Boronat A (2001) FEBS Lett 488:170
141. Seemann M, Bui BT, Wolff M, Tritsch D, Campos N, Boronat A, Marquet A, Rohmer M (2002) Angew Chem Int Edit 41:4337
142. Altincicek B, Duin EC, Reichenberg A, Hedderich R, Kollas AK, Hintz M, Wagner S, Wiesner J, Beck E, Jomaa H (2002) FEBS Lett 532:437
143. Eberl M, Hintz M, Reichenberg A, Kollas AK, Wiesner J, Jomaa H (2003) FEBS Lett 544:4
144. Wolff M, Seemann M, Tse Sum BB, Frapart Y, Tritsch D, Garcia EA, Rodríguez-Concepción M, Boronat A, Marquet A, Rohmer M (2003) FEBS Lett 541:115
145. Brandt W, Dessoy MA, Fulhorst M, Gao W, Zenk MH, Wessjohann LA (2004) ChemBiochem 5:311
146. Hahn FM, Hurlburt AP, Poulter CD (1999) J Bacteriol 181:4499
147. Cunningham FX, Lafond TP, Gantt E (2000) J Bacteriol 182:5841

148. Kajiwara S, Fraser PD, Kondo K, Misawa N (1997) Biochem J 324:421
149. Wang C-W, Oh M-K, Liao JC (1999) Biotechnol Bioeng 62:235
150. Hoeffler J-F, Hemmerlin A, Grosdemange-Billiard C, Bach TJ, Rohmer M (2002) Biochem J 366:573
151. Kuzuyama T, Takagi M, Takahashi S, Seto H (2000) J Bacteriol 182:891
152. Yajima S, Nonaka T, Kuzuyama T, Seto H, Ohsawa K (2002) J Biochem (Tokyo) 131:313
153. Reuter K, Sanderbrand S, Jomaa H, Wiesner J, Steinbrecher I, Beck E, Hintz M, Klebe G, Stubbs MT (2002) J Biol Chem 277:5378
154. Grolle S, Bringer-Meyer S, Sahm H (2000) FEMS Microbiol Lett 191:131
155. Koppisch AT, Fox DT, Blagg BS, Poulter CD (2002) Biochemistry 41:236
156. Richard SB, Bowman ME, Kwiatkowski W, Kang I, Chow C, Lillo AM, Cane DE, Noel JP (2001) Nat Struct Biol 8:641
157. Kemp LE, Bond CS, Hunter WN (2001) Acta Crystallogr D Biol Crystallogr 57:1189
158. Rohdich F, Wungsintaweekul J, Lüttgen H, Fischer M, Eisenreich W, Schuhr CA, Fellermeier M, Schramek N, Zenk MH, Bacher A (2000) Proc Natl Acad Sci USA 97:8251
159. Richard SB, Ferrer JL, Bowman ME, Lillo AM, Tetzlaff CN, Cane DE, Noel JP (2002) J Biol Chem 277:8667
160. Kishida H, Wada T, Unzai S, Kuzuyama T, Takagi M, Terada T, Shirouzu M, Yokoyama S, Tame JR, Park SY (2003) Acta Crystallogr D Biol Crystallogr 59:23
161. Rohdich F, Zepeck F, Adam P, Hecht S, Kaiser J, Laupitz R, Grawert T, Amslinger S, Eisenreich W, Bacher A, Arigoni D (2003) Proc Natl Acad Sci USA 100:1586
162. Kollas AK, Duin EC, Eberl M, Altincicek B, Hintz M, Reichenberg A, Henschker D, Henne A, Steinbrecher I, Ostrovsky DN, Hedderich R, Beck E, Jomaa H, Wiesner J (2002) FEBS Lett 532:432
163. Marshall JH, Wilmoth GJ (1981) J Bacteriol 147:900
164. Johnson EA, Schroeder WA (1996) Adv Biochem Eng Biotechnol 53:119
165. Misawa N, Nakagawa M, Kobayashi K, Yamano S, Izawa Y, Nakamura K, Harashima K (1990) J Bacteriol 172:6704
166. Lee PC, Schmidt-Dannert C (2002) Appl Microbiol Biot 60:1
167. Schmidt-Dannert C (2000) Curr Opin Biotechnol 11:255
168. Arthington-Skaggs BA, Crowell DN, Yang H, Sturley SL, Bard M (1996) FEBS Lett 392:161
169. Yamano S, Ishii T, Nakagawa M, Ikenaga H, Misawa N (1994) Biosci Biotechnol Biochem 58:1112
170. Miura Y, Kondo K, Shimada H, Saito T, Nakamura K, Misawa N (1998) Biotechnol Bioeng 58:306
171. Miura Y, Kondo K, Saito T, Shimada H, Fraser PD, Misawa N (1998) Appl Environ Microbiol 64:1226
172. Shimada H, Kondo K, Fraser PD, Miura Y, Saito T, Misawa N (1998) Appl Environ Microbiol 64: 2676
173. Huang Q, Roessner CA, Croteau R, Scott AI (2001) Bioorg Med Chem 9:2237
174. Reiling KK, Yoshikuni Y, Martin VJJ, Newman J, Bohlmann J, Keasling JD (2004) Biotechnol Bioeng 87:200
175. Martin VJJ, Yoshikuni Y, Keasling JD (2001) Biotechnol Bioeng 75:497
176. Martin VJJ, Pitera DJ, Withers ST, Newman JD, Keasling JD (2003) Nat Biotechnol 21:796
177. Wang C-W, Oh M-K, Liao JC (2000) Biotechnol Prog 16:922
178. Matthews PD, Wurtzel ET (2000) Appl Microbiol Biot 53:396

179. Harker M, Bramley PM (1999) FEBS Lett 448:115
180. Kim S-W, Keasling JD (2001) Biotechnol Bioeng 72:408
181. Albrecht M, Misawa N, Sandmann G (1999) Biotechnol Lett 21:791
182. Kingston DGI (2001) Chem Commun 1:867
183. Sandmann G (2001) Trends Plant Sci 6:14
184. Ruther A, Misawa N, Böger P, Sandmann G (1997) Appl Microbiol Biot 48:162
185. Farmer WR, Liao JC (2001) Biotechnol Prog 17:57
186. Wang G-Y, Keasling JD (2002) Metab Eng 4:193
187. Jackson BE, Hart-Wells EA, Matsuda SPT (2003) Org Lett 5:1629
188. Tapiero H, Townsend DM, Tew KD (2004) Biomed Pharmacother 58:100
189. Nishino H (1998) Mutat Res 402:159
190. Johnson EJ (2002) Nutr Clin Care 5:56
191. Cooper DA, Eldridge AL, Peters JC (1999) Nutr Rev 57:201
192. Albrecht M, Takaichi S, Misawa N, Schnurr G, Böger P, Sandmann G (1997) J Biotechnol 58:177
193. Albrecht M, Takaichi S, Steiger S, Wang Z-Y, Sandmann G (2000) Nat Biotechnol 18:843
194. Yokoyama A, Shizuri Y, Misawa N (1998) Tetrahedron Lett 39:3709
195. Steiger S, Takaichi S, Sandmann G (2002) J Biotechnol 97:51
196. Schmidt-Dannert C, Umeno D, Arnold FH (2000) Nat Biotechnol 18:750
197. Carter OA, Peters RJ, Croteau R (2003) Phytochem 64:425
198. Misawa N, Yamano S, Ikenaga H (1991) Appl Environ Microbiol 57:1847
199. Edwards JS, Palsson BØ (2000) Proc Natl Acad Sci USA 97:5528
200. Reed JL, Vo TD, Schilling CH, Palsson BØ (2003) Genome Biol 4:54
201. Förster J, Famili I, Fu P, Palsson BØ, Nielsen J (2003) Genome Res 13:244
202. Patil KR, Åkesson M, Nielsen J (2004) Curr Opin Biotechnol 15:64

Adv Biochem Engin/Biotechnol (2005) 100: 53–88
DOI 10.1007/b136412

Published online: 5 July 2005

Plant Cells: Secondary Metabolite Heterogeneity and Its Manipulation

Jian-Jiang Zhong[1] (✉) · Cai-Jun Yue[1,2]

[1]State Key Laboratory of Bioreactor Engineering, East China University of Science and Technology, 200237 Shanghai, P.R. China
jjzhong@ecust.edu.cn

[2]College of Life Science and Biotechnology, Heilongjiang August First Land Reclamation University, 163319 Daqing, P.R. China

Abstract This chapter proposes the concept of rational manipulation of secondary metabolite heterogeneity in plant cell cultures. The heterogeneity of plant secondary metabolites is a very interesting and important issue because these structure-similar natural products have different biological activities. Both taxoids and ginsenosides are two kinds of preeminent examples in the enormous reservoir of pharmacologically valuable heterogeneous molecules in the plant kingdom. They are derived from the five-carbon precursor isopentenyl diphosphate, produced via the mevalonate or the non-mevalonate pathway. The diterpenoid backbone of taxoids is synthesized by taxadiene synthase and the triterpenoid backbone of ginsenosides is synthesized by dammarenediol synthase or β-amyrin synthase. After various chemical decorations (oxidation, substitution, acylation, glycosylation, benzoylation, and so on) mediated by P450-dependent monooxygenases, glycosyltransferases, acyltransferases, benzoyltransferases, and other enzymes, the terpenoid backbones are converted into heterogeneous taxoids and ginsenosides with different bioactivities. Although detailed information about accumulation and regulation of individual taxoids or ginsenosides in plant cells is still lacking, remarkable progress has recently been made in the structure and bioactivity identification, biosynthetic pathway, manipulation of their heterogeneity by various methodologies including environmental factors, biotransformation, and metabolic engineering in cell/tissue cultures or in plants. Perspectives on a more rational and efficient process to manipulate production of desired plant secondary metabolites by means of metabolic engineering and "omics"-based approaches (e.g., functional genomics) are also discussed.

Keywords Plant cell · Heterogeneity · *Taxus* spp. · Ginseng · Manipulation · Secondary metabolite

1 Introduction

Higher plants, about 400 000 species in the world [1], are a valuable source of numerous metabolites, which are used as pharmaceuticals, agrochemicals, flavors, fragrances, colors, biopesticides, and food additives. More than 100 000 plant secondary metabolites have already been identified, which probably represent only 10% of the actual total in nature and only half the structures have been fully elucidated [2–4]. Molecular diversity is a widely existing phenomenon in nature, and many plant secondary metabolites are structure-similar but bioactivity-different. The enormous heterogeneity of plant secondary metabolites is usually derived from differential modification of common backbone structures. For example, over 5000 different flavonoids and 300 different glycosides of a single flavonol, quercetin, have already been identified [5]. The immense diversity of plant secondary metabolites is often obtained by derivatization of specific lead structures through post-biosynthetic events such as hydroxylation, glycosylation, methylation, acylation, prenylation, sulfation, and benzoylation [6]. Hundreds of secondary metabolite modifying enzymes (e.g., oxidases, acyltransferases, methyltransferases, glycosyltransferases, sulfotransferases, and benzoyltransferase) have been cloned and characterized [7, 8].

Generally, the function of each plant secondary metabolite is different. Figure 1 shows terpenoids as an extremely fascinating example; they are present in all organisms but are especially abundant in plants, with more than 30 000 compounds reported to date [9–11]. Terpenoids are the most functionally and structurally diverse group of plant natural products that include diterpenoid alkaloids, sterols, triterpene saponins, and related structures. The most basic function of triterpenes is to give membranes stability, such as β-sitosterol (**1** in Fig. 1) does in plants. By further oxygenation, for example, castasterone (**2** in Fig. 1), acts as signals that interfere with morphological differentiation in plants. Furthermore, triterpene glycosides, such as saponin phytoalexins (**3** in Fig. 1), damage fungal membranes by significantly reducing their stability [12].

Many structure-similar but bioactivity-different secondary metabolites are usually generated in one plant. Both taxoids (diterpenoid alkaloids originally isolated from the bark of the Pacific yew, *Taxus brevifolia*) and ginseng saponins (ginsenoside, an active group of triterpene saponins mostly from

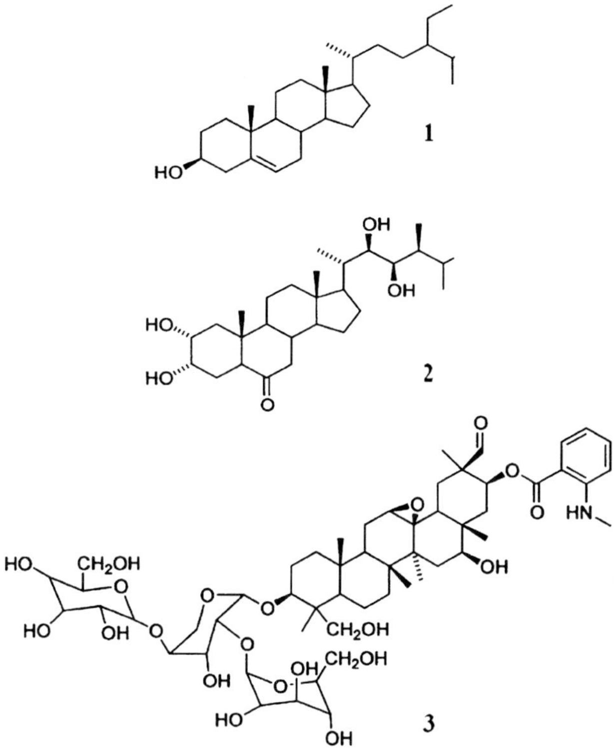

Fig. 1 Triterpenes with diverse biological activities: β-sitosterol (**1**) confers membrane stability in plants; castasterone (**2**), a brassinosteroid growth hormone; avenacin A-1 (**3**), antifungal saponin phytoalexin. Refer to the text for details

Panax ginseng, *P. notoginseng* or *P. quinquefolium*) are tremendously heterogeneous. Anticancer potency of each taxoid is different [13]. The biological activities of some ginsenosides even oppose each other. For example, Rg_1 has the effect of stimulating the central nervous system, whereas Rb_1 has tranquilizing effects on the central nervous system and Rc inhibits the central nervous system [14, 15]. However it is difficult to manipulate their heterogeneity in field-cultivated plants; therefore, the pharmacodynamic instability of these herbs often takes place owing to the change of the quality of the raw materials (especially in both the composition and the distribution of related metabolites). The purification of an individual compound is a current approach for maintaining certain specific potency, but the metabolite (taxoid and ginsenoside) content is usually quite low, while the physicochemical characteristics of various analogues (taxoids or ginsenosides) are very similar; therefore, their separation and purification is an expensive and very complicated process, and the yields of active compounds from plants are season- and environment-dependent. Cell and tissue culture is an attractive alternative source to a whole plant for production of the high-value-added secondary metabolites. This chapter proposes the concept of rational manipulation of secondary metabolite heterogeneity in plant cell cultures. It is very advantageous to intentionally manipulate the heterogeneity of secondary metabolites in plant cell and tissue cultures by altering or stimulating their genome and/or the subsequent processes, which result in the desired enzymatic syntheses of secondary metabolites. The manipulating techniques utilized include elicitation, hormone treatment, enzyme inhibition, growth-retardant treatment, and precursor-directed biosynthesis resulting in the production of previously undiscovered plant metabolites or a change of the production ratio of certain secondary metabolites [16]. Of course, other engineering strategies, such as temperature shift and change of oxygen partial pressure, also affect the heterogeneity of plant secondary metabolites in cell cultures. Biotransformation by various organisms and enzymes is an effective method for changing the heterogeneity of plant secondary metabolites. Metabolic engineering approaches are promising in manipulating the accumulation of plant secondary metabolites. In the following, by taking taxoid and ginsenoside as typical examples, progress in the structure and activity identification, biosynthesis, and manipulation of their heterogeneity in plants, their tissues or cells is reviewed.

2 Heterogeneity of Taxoid and Its Manipulation

2.1 Taxoid and Its Diversity

Taxoids are complex, substituted diterpenoids, one of which, the famous taxol (paclitaxel), was first isolated from the bark of *T. brevifolia* Nutt and its structure was defined in 1971 [17]. Subsequently, paclitaxel and taxoid derivatives have been reported from foliage and bark of several other species of *Taxus*, like *T. wallichinan*, *T. baccata*, *T. canadensis*, *T. cuspidata*, and *T. yunnanesis* [18–22]. In addition to the plant source, some endophytic fungi, such as *Tubercularia* sp., *Sporormia minima*, and *Seimatoantlerium tepuiense*, have also been reported to produce taxol and other taxoids [23–25].

Until now, over 350 taxoids have been classified into 16 groups (Table 1) [26]. Chemical derivatization of taxoids contributes to the diversity of taxoid function. Taxoids are well-known antineoplastic drugs, and are used to treat a range of cancers, either alone or in combination with other chemotherapeutic agents [27, 28]. Guéritte [29] summarized the general structure-antitubulin activity relationship (Fig. 2). Paclitaxel is a highly functionalized taxoid that acts by promoting tubulin polymerization, ultimately leading to

Table 1 Classification of taxoids

Class	Structure
Neutral taxoids with a C-4(20) double bond	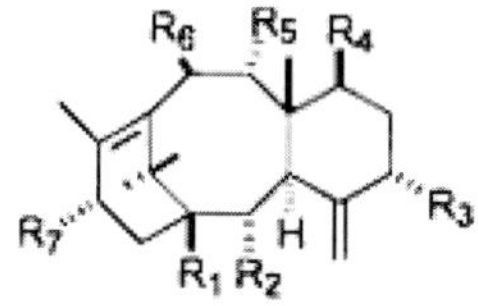
Basic taxoids with a C-4(20) double bond	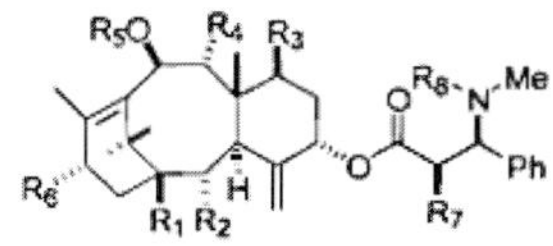
5-Cinnamoyl taxoids with a C-4(20) double bond	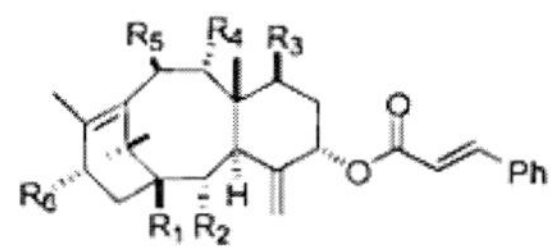
Taxoids with a C-4(20) double bond and oxygenation at C-14	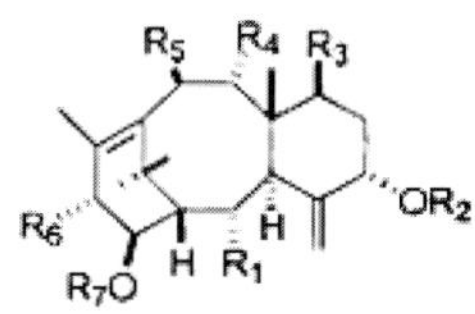
Taxoids with a C-12(16)-oxido bridge and a C-4(20)double bond	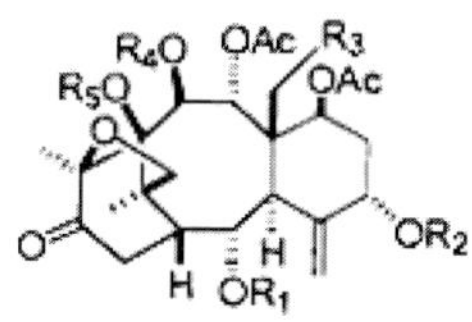
Taxoids with a C-4(20) epoxide	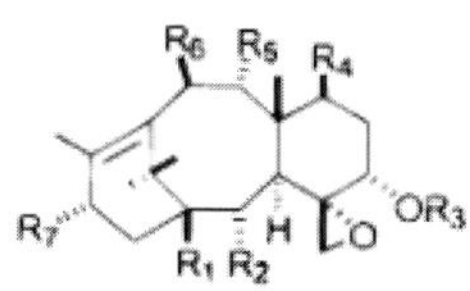
Taxoids with an oxetane ring	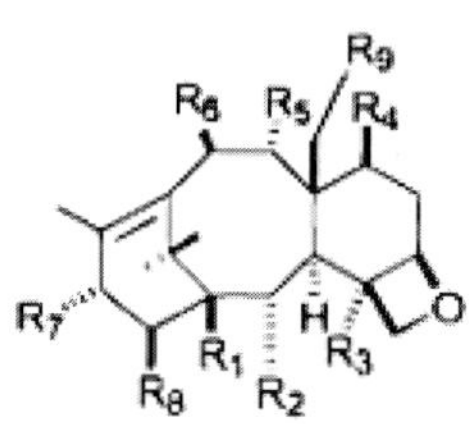

Table 1 (continued)

Class	Structure
Taxoids with an oxetane ring and a phenylisoserine C-13 side chain	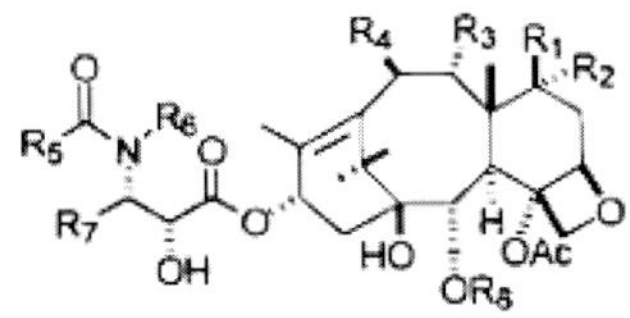
Taxoids with an open oxetane or oxirane ring	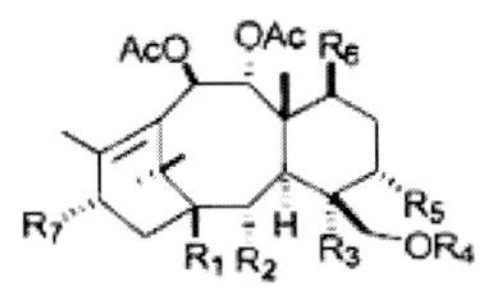
11(15f1)-*abeo*-Taxoids with a C-4(20) double bond	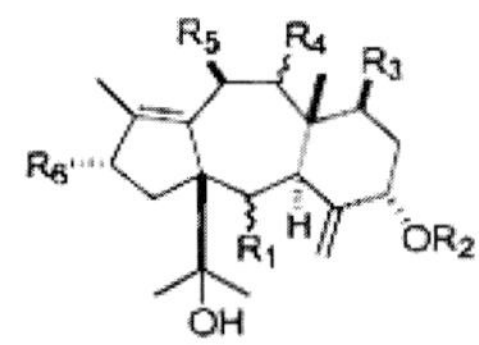
11(15f1)-*abeo*-Taxoids with an oxetane ring	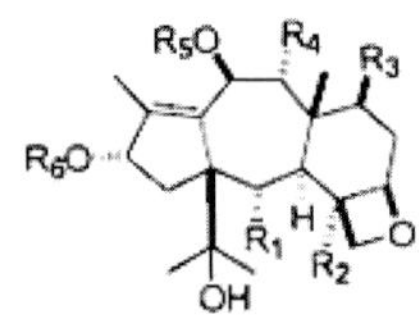
11(15f1)-*abeo*-Taxoids with an open oxetane or oxirane ring	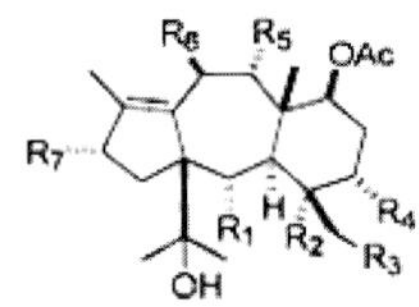
3,8-*seco*-Taxoids	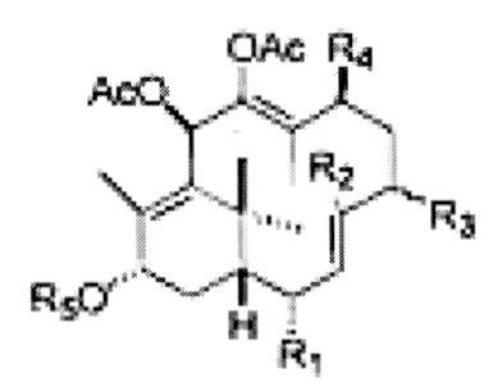

cell death [30]. The structural elements (pharmacophores) responsible for the cytotoxicity of paclitaxel, in addition to the rigid taxane skeleton, include the oxetane ring (D-ring), the *N*-benzoylphenylisoserine side chain appended to C-13, the benzoate group at C-2, and the acetate function at C-4 of the taxane ring [31]. In 120 taxoids isolated from the Japanese yew, *T. cuspidate*, only four non-paclitaxel-type taxoids (taxuspine D, taxezopidines K and L, and

Table 1 (continued)

Class	Structure
Taxoids with a C-3(11) bridge and a C-4(20) double bond	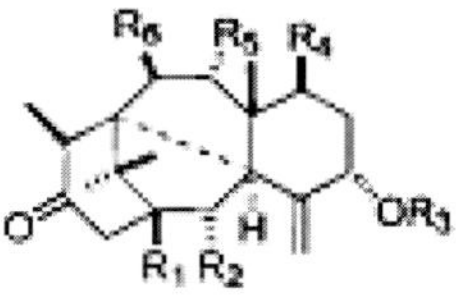
2(3*f*20)-*abeo*-Taxanes	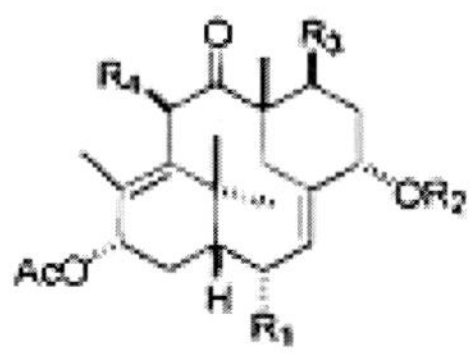
Other miscellaneous taxoids	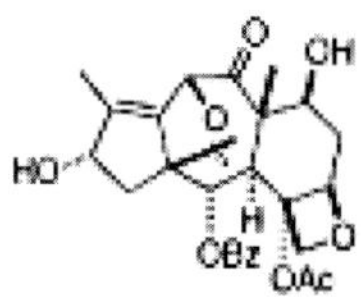 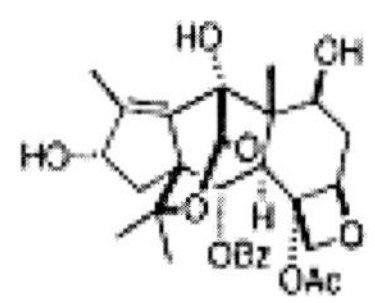

taxagifine) exhibit potent inhibitory activity against Ca^{2+}-induced depolymerization of microtubules, while taxuspine D induces spindles with strong birefringence in the same manner as paclitaxel [32].

2.2 Taxoid Biosynthesis and Manipulation of Taxoid Heterogeneity

2.2.1 Taxoid Biosynthesis

A typical biosynthetic pathway of taxoids, by taking paclitaxel as an example, is illustrated in Fig. 3. The diterpenoid skeleton of taxoids, as with other terpenoids of plastid origin, was observed by using labeling studies with ^{13}C-labeled glucose to be derived via the 1-deoxy-D-xylulose-5-phosphate pathway [33–37], in which the isopentenyl diphosphate formed is employed in the biosynthesis of carotenoids, phytol, plastoquinone, isoprene, monoterpenes, and diterpenes. The committed step in the biosynthesis of paclitaxel and other taxoids is represented by the cyclization of the universal diterpenoid precursor geranylgeranyl diphosphate (GGPP) to taxa-4(5),11(12)-diene [38]. Taxadiene synthase, a 79-kDa diterpene cyclase, catalyzes this reaction, which is slow but apparently not rate-limiting [39, 40].

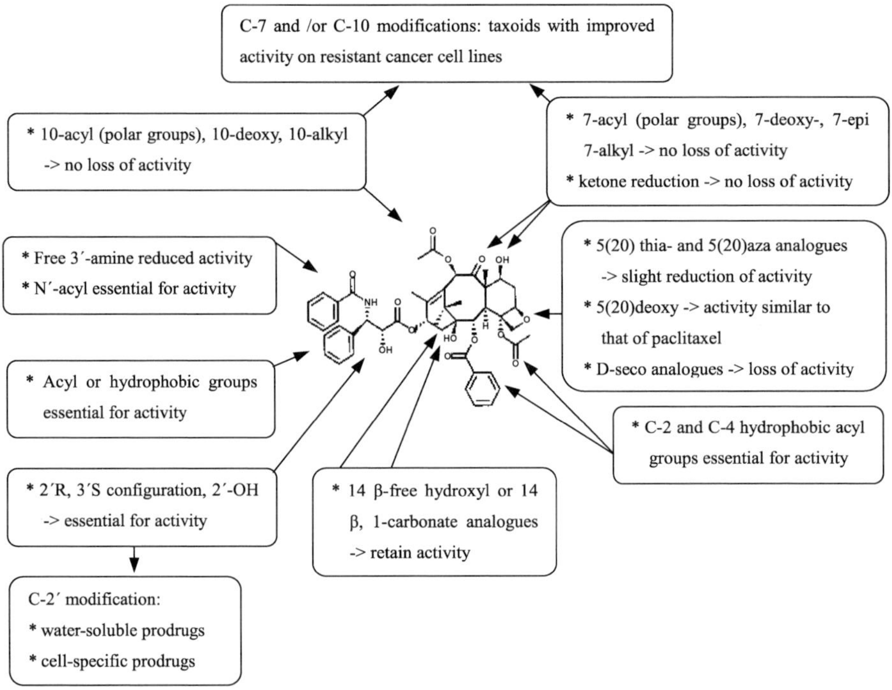

Fig. 2 The general structure–antitubulin activity relationships of taxoids (modified from the literature [29])

On the other hand, the enzyme was demonstrated to be a key one in the biosynthesis of a taxoid, taxuyunnanine C ($2\alpha,5\alpha,10\beta,14\beta$-tetraacetoxy-4(20),11-taxadiene, Tc), by suspended cells of *T. chinensis* in response to methyl jasmonate (MJA) elicitation [41]. The second specific step in taxoid biosynthesis is considered to be the cytochrome P450 dependent hydroxylation at the C-5 position of the taxane ring, which is accomplished by allylic rearrangement of the 4(5) double bond to the 4(20) position to yield taxa-4(20),11(12)-diene-5α-ol [42]. Taxa-4(20),11(12)-diene-5α-ol is a branching point in the paclitaxel pathway to form other naturally occurring taxanes. The enzymes taxadien 13α-hydroxylase and taxadien-5α-ol acetyltransferase, which catalyze taxa-4(20),11(12)-diene-5α-ol to produce different taxoids, were reported [43, 44]. Taxadiene-5α-10β-diol monoacetate was another possible branching point in the paclitaxel pathway. It can be transformed into 5α-acetoxy-$10\beta,14\beta$-dihydroxy taxadiene by taxoid 14β-hydroxylase, but it is still not known how it is transformed into 2-debenzoyltaxane or taxasin [45, 46]. However, previous evaluations [47] of the relative abundance of naturally occurring taxanes [26, 48] have suggested that hydroxylations at positions C-5, C-10, C-9, and C-2 are earlier than that at positions C-13, C-1, and C-7 of the taxane ring in pacli-

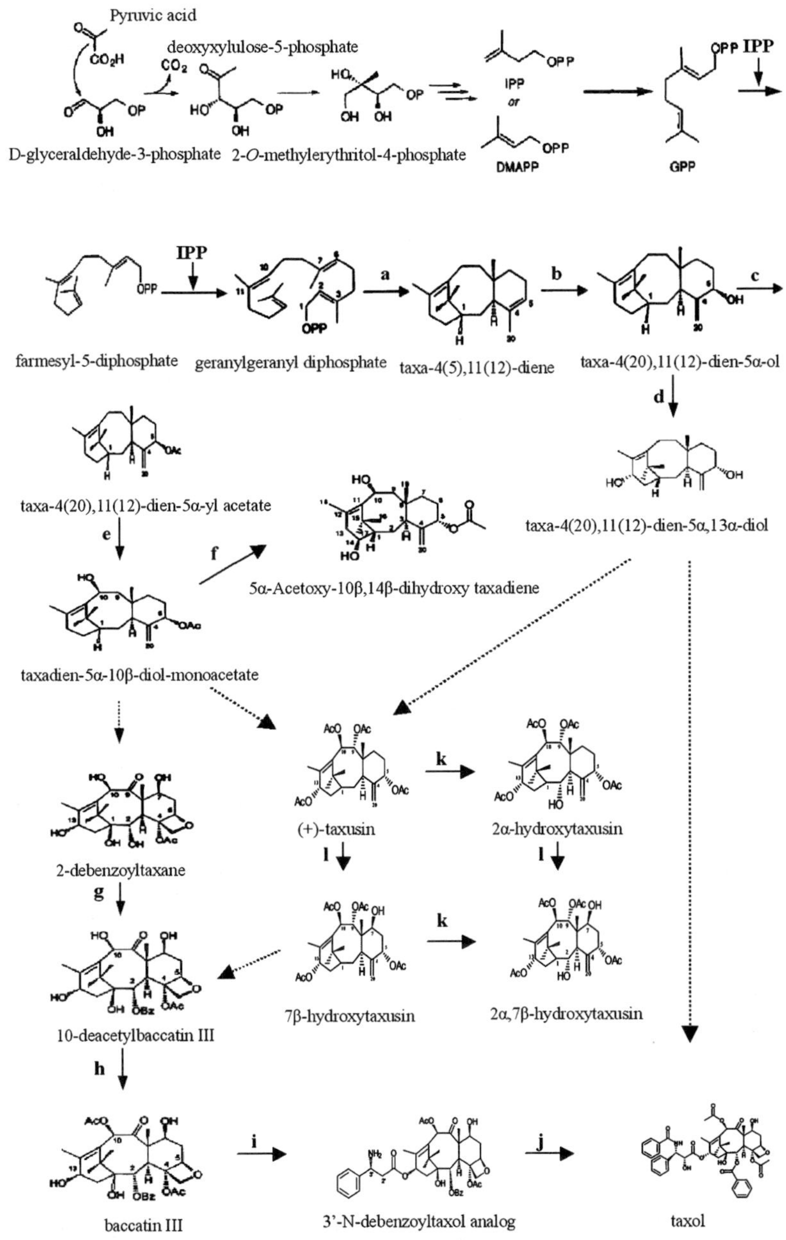

Fig. 3 The proposed paclitaxel biosynthetic pathway. The enzymes indicated are *a* taxadiene synthase, *b* taxadiene 5α-hydroxylase, *c* taxadien-5α-ol acetyltransferase, *d* taxadien 13α-hydroxylase, *e* 10α-hydroxylase, *f* 14β-hydroxylase, *g* 2α-*O*-benzoyltransferase, *h* 10-*O*-acetyltransferase, *i* phenylpropanoyltransferase, *j* 3′-*N*-debenzoyl-2′-deoxytaxol *N*-benzoyltransferase, *k* 7β-hydroxylase, and *l* 2α-hydroxylase. The *broken arrow* indicates multiple convergent steps (modified from Refs. [43–46, 51–54])

taxel biosynthesis, and several biosynthetic mechanisms have been proposed for formation of the oxetane ring (D-ring) [49, 50]. Taxusin, a presumed dead-end metabolite of yew heartwood, may also be from taxa-4(20),11(12)-dien5α,13β-diol and/or taxadiene-5α-10β-diol monoacetate, although the details are unclear. Taxusin is another node in the biosynthesis of taxoids, and can efficiently be converted to the corresponding 2α–hydroxytaxusin and 7β-hydroxytaxusin by the taxoid 2α-hydroxylase and the taxoid 7β-hydroxylase, respectively. It is also possible that 7β-hydroxytaxusin will be converted to 2-debenzoyltaxane [46]. Until now the pathway from 2-debenzoyltaxane to paclitaxel has been clear, and includes the formation of 2-benzoxy taxoid by taxane 2α-*O*-benzoyltransferase, the conversion of 10-deacetylbaccatin III to baccatin III by 10-*O*-acetyltransferase, side-chain attachment by the phenylpropanoyltransferase, and side-chain benzamidation by 3′-*N*-debenzoyl-2′-deoxytaxol *N*-benzoyltransferase to form paclitaxel [51]. Given the very large number of structurally defined taxoids, and that there are even multiple pathways from taxadiene to paclitaxel, there must also exist several side routes and diversions responsible for the formation of various taxoids. The substrate selectivities of the taxoid hydroxylases and acyltransferases almost certainly play a central role in the formation of heterogeneous taxoids.

2.2.2 Manipulation of Taxoid Heterogeneity

Since paclitaxel has been found to exhibit significant antitumor activity against various cancers, and there is poor availability of paclitaxel from natural sources (only 50–150 mg/kg of dried trunk bark can be isolated from several species of yew), great attention has been paid to other supply sources. Except for semisynthesis from its natural precursor 10-deacetylbaccatin III, which is mainly obtained from leaves of *Taxus* species, plant cell and tissue culture of *Taxus* species is considered as one of the most promising approaches to obtain paclitaxel and related taxoids. It is practical to manipulate taxoid heterogeneity in cell cultures via environmental factors and molecular biology techniques.

2.2.2.1 Effect of Temperature Shift

Biosynthesis of taxoids in cultured *Taxus* cells was affected by temperature shift during cultivation. When the temperature was shifted from 24 to 29 °C at day 21 in cell cultures of *T. chinensis* treated with 4 μM silver nitrate at the initial cultivation time, the yield of paclitaxel increased from 49.6 to 82.4 mg/L at day 35, while that of Tc decreased from 885.9 to 512.9 mg/L [55]. The results imply that the biosyntheses of different taxoids might have their own temperature preference, and the temperature-shifting strategy to produce a specific taxoid by cultured cells should be varied accordingly.

2.2.2.2 Effect of Methyl Jasmonate

New taxoids may be produced or primary taxoids lost in cultured *Taxus* cells after elicitation with MJA, a key signal compound which is widely used in the production of secondary metabolites by plant cells. In the CR-5 callus culture of *T. cuspidate* [56], it is reported that after stimulation with 100 μM MJA, five more taxoids, cephalomannine, 1β-dehydroxybaccatin VI, taxinine NN-11, baccatin I, and 2α-acetoxytaxusin, and one more abietane, taxamairin C, were produced in addition to known taxoids, paclitaxel, 7-*epi*-taxol, taxol C, baccatin VI, taxayuntin C, taxuyunnanine C and its analogues, and yunnanxane, and an abietane, taxamairin A. After 60-days elicited cultivation, the levels of taxuyunnanine C and its analogues increased 3.1-fold, and paclitaxel and its analogues increased 5.2-fold compared with those in CR-5 without MJA elicitation. The production of phenolic abietane derivatives, taxamairin A and taxamairin C, was promoted a little [56]. Ketchum et al. [57] reported that after MJA elicitation Mh00D cell lines of *T.* x *media* cv. *Hicksii* produced a new taxoid, 1β-dehydroxybaccatin VI, and lost baccatin III and 10-deacetylbaccatin III, but Mh00W cell lines of *T.* x *media* cv. *Hicksii* produced new taxoids, 1β-dehydroxybaccatin VI, baccatin III, and 5α,7β,9α,10β,13α-pentaacetoxy-2*a*-benzoyloxytaxa-4(20),11-diene, and lost baccatin VI. These results imply that MJA altered the heterogeneity of taxoids by activating certain pathways of taxoid synthesis and/or reducing certain primary pathways in different cell lines. It is necessary to have the metabolic and physiological characterization of cell lines while manipulating the heterogeneity of the products.

In *T. canadensis* (CO93P) suspension cultures with or without 200 mM MJA elicitation, the distribution of taxoids was similar [58]. All of the major taxoids present in the elicited cultures were also present in the nonelicited cultures, but the relative proportion of the taxoids was different. These observations may indicate that MJA elicitation affects the relative abundance of existing taxoids in certain *Taxus* species, even if elicitation does not result in the production of novel taxoids. This may be caused by the accumulation of intermediates as a result of one or more rate-limiting steps in the taxoid biosynthetic pathway.

2.2.2.3 Effect of Precursors, Growth Retardants, and Phenylalanine Ammonia Lyase Inhibitors

Veeresharm et al. [59] reported that precursors and growth retardants showed different improvement of the production of paclitaxel, deacetylbaccatin III, and baccatin III in *T. wallichiana* cell cultures (Fig. 4). The accumulation of deacetylbaccatin III, baccatin III, or paclitaxel enhanced by addition of the precursors phenylalanine (1 mM), sodium benzoate (0.2 mM), hippuric acid (1 mM), and leucine (1 mM) was different in cell cultures. Hippuric

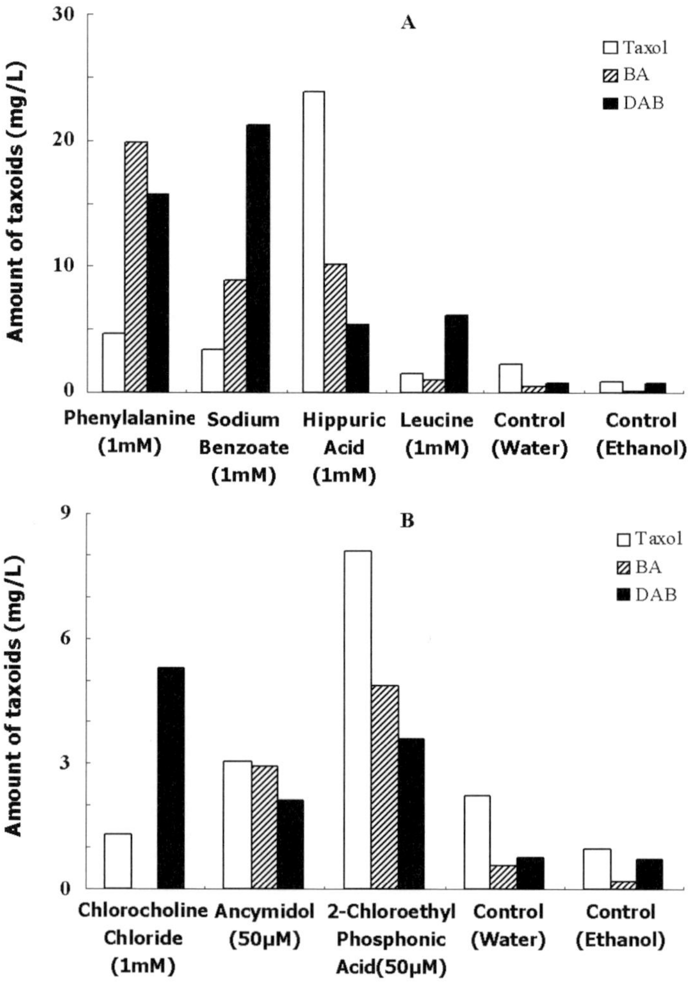

Fig. 4 Effect of **a** precursors and **b** growth retardants on taxoid production in cell cultures of *Taxus wallichiana* (modified from Ref. [56])

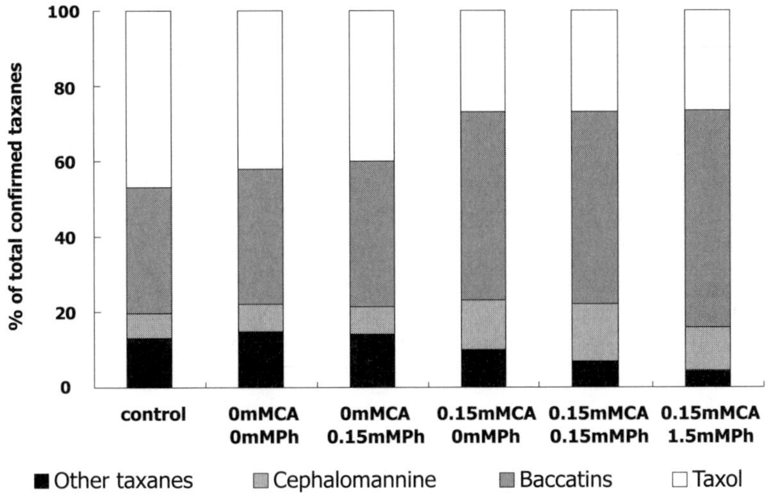

Fig. 5 Single or combined addition of cinnamic acid (*CA*, 0.15 mM) and phenylalanine (*Ph*, 0.15 and 1.5 mM) to CO93P *T. canadensis* cultures at day 7. Taxoids were measured at day 15. The baccatins consist of greater than 96% 13-acetyl-9-dihydrobaccatin III and 9-dihydrobaccatin III (modified from Ref. [57])

acid was most favorable for accumulation of paclitaxel, sodium benzoate for baccatin III, and phenylalanine for deacetylbaccatin III. Like precursors, growth retardants 2-chloroethyl phosphonic acid (50 μM) and chlorocholine chloride (1 mM) were beneficial to the production of paclitaxel and deacetylbaccatin III, respectively. This may be due to the different response of 2α-*O*-benzoyltransferase, 10-*O*-acetyltransferase, phenylpropanoyltransferase, and 3′-*N*-debenzoyl-2′-deoxytaxol *N*-benzoyltransferase to these precursors and growth retardants. These precursors and growth retardants can be potential regulators of the taxoid heterogeneity.

Brincat et al. [60] reported the effect of cinnamic acid (a phenylalanine ammonia lyase, PAL, inhibitor) and phenylalanine on the synthesis of total taxanes in CO93P *T. canadensis* cultures (Fig. 5). The concentration of 13-acetyl-9-dihydrobaccatin III and 9-dihydrobaccatin III at least doubled in CO93P cells treated with 0.15 mM cinnamic acid, although phenylalanine had very little effect on the taxane profile. Considering α-aminooxyacetic acid (a PAL inhibitor), which almost entirely shut down paclitaxel production, and L-α-aminooxy-β-phenylpropionic acid (another PAL inhibitor), which slightly enhanced paclitaxel production, they suggested that the impact of cinnamic acid on paclitaxel might be related not to its effect on PAL but rather to a specific effect on the taxane pathway.

2.2.2.4 Biotransformation

Biotransformation is a biosynthetic or degradation process using enzymes in living organisms or isolated from living cells as biocatalysts. The characteristics of biotransformation are regioselective and stereoselective reaction under mild conditions and easy production of optically active compounds. It is one of the methodologies to produce diverse taxoids. The investigation of biotransformation of taxoids is gaining more and more interest, with their reactions performed by bacteria, fungi, plant cells, and isolated enzymes. Hydroxylation, acylation, epoxidation, hydrolysis, recomposition, and other reactions are generated in biotransformation of taxoids. For example, sinenxan A (a taxoid) can be easily transformed by many organisms (Fig. 6, Table 2). Taxoids can also be transformed directly by various cell-free enzymes, which are very useful in manipulation of taxoid heterogeneity. Patel [68] reported that C-13 taxolase (which catalyzes the cleavage of the C-13 side chain of various taxanes) derived from *Nocardioides albus* SC 13911, C-10 deacetylase (which catalyzes the cleavage of C-10 acetate of various taxanes) derived from *N. luteus* SC 13912, and C-7 xylosidase (which catalyzes the cleavage of C-7 xylose from various xylosyltaxanes) derived from *Morexella* sp. SC 13963 converted various taxanes in extracts of *Taxus* cultivars to 10-deacetylbaccatin III, whose concentration was increased by 5.5- to 24-fold. The C-10 deacetylase also can transform 10-deacetylbaccatin III to baccatin III with a reaction yield of 51% [69]. Recently, conversion from 7-deoxy-10-deacetylbaccatin III into 6-hydroxy-7-deoxy-10-deacetylbaccatin III by *N. luteus* SC 13912 (ATCC 55426) was reported [70].

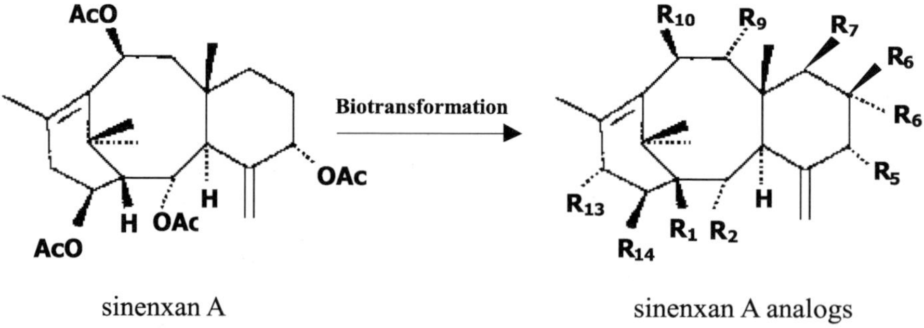

Fig. 6 Biotransformation of sinenxan A by various organisms. The *R* groups and biocatalysts are shown in Table 2

Table 2 Biotransformation of sinenxan A by various organisms

Structures of products	Species of organisms
$R_5 = OH, R_2 = R_{10} = R_{14} = AcO$	*Catharanthus roseus* [61]
$R_{10} = OH, R_2 = R_5 = R_{14} = AcO$	*Platycodon grandiflorum* [61, 62]
$R_1 = OH, R_5 = R_9 = R_{10} = R_{13} = AcO$	*Absidia coerulea* [63]
$R_{14} = OH, R_5 = R_9 = R_{10} = R_{13} = AcO$	*A. coerulea* [63]
$R_5 = R_{10} = R_{14} = OH, R_2 = AcO$	*Cunninghamella echinulata* [64]
$R_5 = R_6 = R_{10} = R_{14} = OH, R_2 = AcO$	*C. elegans* [64]
$R_5 = R_6 = R_{10} = OH, R_2 = R_{14} = AcO$	*C. echinulata* [64]
$R_5 = R_{10} = OH, R_2 = R_6 = R_{14} = AcO$	*C. echinulata* [64]
$R_6 = R_{10} = OH, R_2 = R_5 = R_{14} = AcO$	*C. roseus, C. echinulata, Ginkgo biloba* [61, 65, 66]
$R_6 = R_9 = R_{10} = OH, R_2 = R_5 = R_{14} = AcO$	*C. roseus, G. biloba* [61, 66]
$R_6 = OH, R_2 = R_5 = R_{10} = R_{14} = AcO$	*C. elegans* [64]
$R'_6 = R_{10} = OH, R_2 = R_5 = R_{14} = AcO$	*C. echinulata* [67]
$R_7 = OH, R_2 = R_5 = R_{10} = R_{14} = AcO$	*A. coerulea* [63]
$R_9 = R_{10} = OH, R_2 = R_5 = R_{14} = AcO$	*G. biloba* [66]
$R_9 = R_{14} = OH, R_2 = R_5 = R_{10} = AcO$	*G. biloba* [66]
$R_9 = OH, R_2 = R_5 = R_{10} = R_{14} = AcO$	*G. biloba* [66]
$R_9 = OCHO, R_2 = R_5 = R_{10} = R_{14} = AcO$	*G. biloba* [66]
$R_{10} = OCHO, R_2 = R_5 = R_{10} = R_{14} = AcO$	*G. biloba* [66]
$R_6 = R_9 = R_{10} = OH, R_2 = R_5 = R_{14} = AcO$	*C. roseus, G. biloba* [61, 66]

The skeletons of sinenxan A analogs are shown in Fig. 6.

2.2.2.5
Metabolic Engineering Approach

A metabolic engineering approach to engineer cells is a new method for directed production of desired taxoids. It was reported that in *Escherichia coli* cells transformed to express three genes encoding four enzymes of the terpene biosynthetic pathway (including the committed GGPP synthase and taxadiene synthase), taxadiene could be conveniently synthesized in vivo at the unoptimized yield of 1.3 mg/L [71]. Considering a limited pool of precursors to GGPP and the requirement of P450 monooxygenases for further biosynthesis of other taxoids, engineered *E. coli* cells are not better than engineered plant cells; thus, Besumbes et al. [72] reproduced some functional steps of the paclitaxel biosynthetic pathway in *Arabidopsis thaliana* plants to produce taxadiene. A complementary DNA (cDNA) encoding the full-length taxadiene synthase from *T. baccata* was successfully integrated into the *A. thaliana* genome. The constitutive production of the enzyme in *A. thaliana* led to the accumulation of taxadien, and induction of transgene expression using a glucocorticoid-mediated system consistently resulted in a more efficient recruitment of GGPP for the production of taxadiene, which reached a level 30-fold higher than that (around 20 ng/g dry weight) in plants constitutively expressing the transgene.

3 Heterogeneity of Ginsenoside and Its Manipulation

3.1 Ginsenoside and Its Diversity

Ginsenosides are a group of triterpenoid saponins. More than 30 ginsenosides have been isolated from ginseng plants and their chemical structures have been identified. As shown in Table 3, representative ginsenosides exhibit considerable structural variation. In the same type ginsenosides, they differ from one another by the types of sugar moieties, their number, and their site of attachment. Some sugar moieties present are glucose, xylose, rhamnose, and arabinose. They are usually attached to C-3, C-6, or C-20 with formation of chains of a single sugar moiety or oligosaccharide. Ginsenosides also differ in the number and the site of attachment of hydroxyl groups. Compared with that of protopanaxadiol-type ginsenosides, the aglycone of protopanaxatriol-type ginsenosides (protopanaxatriol) has one more hydroxyl group at C-6, which possibly stems from protopanaxadiol by oxidation. Another factor that contributes to structural differences between ginsenosides is the stereochemistry at C-20. Most ginsenosides that have been isolated are naturally present as enantiomeric mixtures [73, 74]. The binding site of the sugar, the number of hydroxyl groups, and the stereoisomerism of ginsenosides have been shown to influence their biological activities.

Numerous reports have been published on the pharmacological and biological activities of various ginsenosides as summarized in Table 4 [75]. There is a very close relationship between the structure and the function of ginsenosides. Both ginsenoside Rd and Rb_1 are protopanaxadiol-type ginsenosides, which differ only by the presence of two glucose moieties at C-20 in Rb_1 and one glucose moiety in Rd. Except for vasodilating action, they do not share the same pharmacological functions (Table 4). Ginsenosides Rh_1 and Rh_2 are also structurally similar. Rh_2 inhibited in vitro proliferation of lung cancer cells 3LL (mice), Morris liver cancer cells (rats), B-16 melanoma cells (mice), and HeLa cells (human) and stimulated melanogenesis and cell-to-cell adhesiveness, but Rh_1 had no effects on cell growth and cell-to-cell adhesiveness despite its stimulation of melanogenesis [76]. Furthermore, only Rh_2 was incorporated in the lipid fraction of the B16–BL6

Table 3 Representative ginsenosides of ginseng congeners

Ginsenoside	R_1	R_2
Protopanaxadiol type		
Rh_2	Glc	H
F_2	Glc	Glc
Rg_3	Glc(2-1)Glc	H
Rd	Glc(2-1)Glc	Glc
Rb_1	Glc(2-1)Glc	Glc(6-1)Glc
Rb_2	Glc(2-1)Glc	Glc(6-1)Arap
Rb_3	Glc(2-1)Glc	Glc(6-1)Xyl
Rc	Glc(2-1)Glc	Glc(6-1)Araf
Ra	Glc(6-1)Glc(6-1)Glc	Glc(3-1)Glc3-1)Glc
Ra_1	Glc(2-1)Glc	Glc(6-1)Arap(4-1)Xyl
Ra_2	Glc(2-1)Glc	Glc(6-1)Arap(2-1)Xyl
Ra_3	Glc(2-1)Glc	Glc(6-1)Arap(3-1)Xyl
Rs_1	Glc(2-1)Glc(6)Ac	Glc(6-1)Arap
Rs_2	Glc(2-1)Glc(6)Ac	Glc(6-1)Araf
Protopanaxatriol type		
Re	Glc(2-1)Rha	Glc
Rf	Glc(2-1)Glc	H
Rg_1	Glc	Glc
Rg_2	Glc(2-1)Rha	H
Rh_1	Glc	H
F_1	H	Glc
F_3	H	Glc(6-1)Arap
Oleanane type		
Ro	Glc(2-1)Glc	Glc

The skeletons of ginsenosides are shown in Fig. 7.
Glc β-D-glucopyranose, *Arap* α-L-arabopyranose, *Araf* α-L-arabofuranose, *Xyl* β-D-xylopyranose, *Rha* α-*l*-rhamnopyranose, *Ac* acetyl

melanoma cell membrane. Differences in the number of hydroxyl groups have also been shown to influence pharmacological activity. Ginsenosides Rh_2 and Rh_3, which possibly stem form protopanaxadiol, are different only by the presence of a hydroxyl group at C-20 in Rh_2. Both Rh_2 and Rh_3 induced the differentiation of promyelocytic leukemia HL-60 cells into morphological and functional granulocytes, but the potency of Rh_2 was higher [77].

Since the modules with which stereoisomers react in biological systems are also optically active, they are considered to be functionally different chemical compounds [78]. Consequently, they often differ considerably in potency, pharmacological activity, and pharmacokinetic profile. Both 20(*S*) and 20(*R*) ginsenoside Rg_2 inhibited acetylcholine-evoked secretion of catecholamines from cultured bovine adrenal chromaffin cells [79]. However, the 20(*S*) isomer showed a greater inhibitory effect. Many factors may contribute to the

Table 4 Pharmacological actions of various ginsenosides

	Ginsenosides
Antiplatelet aggregation	Ro, Rg_1, Rg_2
Fibrinolytic action	Ro, Rb_1, Rb_3, Rc, Re, Rg_1, Rg_2
Stimulation of phagocytic action	Ro, Rb_1, Rb_2, Rc, Rg_3, Rh_2, Re, Rg_2, Rh_1
Vasodilating action	Rb_1, Rd, Rg_1
Cholesterol and neutral lipid decreasing and HDL-cholesterol increasing effects	Rb_1, Rb_2, Rc
Stimulation of ACTH corticosterone secretion	Rb_1, Rb_2, Rc, Re
Stimulation of RNA polymerase, protein synthesis	Rb_1, Rc, Rg_1
Inhibition of cancer cell invasion	Rg_3
Induction of reverse transformation	Rh_2
Inhibition of tumor angiogenesis	Rb_2

multiple pharmacological effects of ginsenosides. The structural isomerism and stereoisomerism exhibited by ginsenosides increase their pharmacological diversity.

3.2 Ginsenoside Biosynthesis and Manipulation of Ginsenoside Heterogeneity

3.2.1 Ginsenoside Biosynthesis

Ginsenosides are synthesized via the isoprenoid pathway by cyclization of 2,3-oxidosqualene to give primarily oleanane dammarane triterpenoid skeletons (dammarenediol or β-amyrin). The first committed step in the synthesis of triterpenoid saponins involves the cyclization of 2,3-oxidosqualene to give one of a number of different potential products. Ginsenosides are derived from dammarane skeletons or oleanane. Dammarenyl cation produced by this cyclization forms a branching point in the ginsenoside biosynthetic pathway (Fig. 7).

The oleanane or dammarane skeleton undergoes various modifications (oxidation, substitution, and glycosylation), mediated by cytochrome P450 dependent monooxygenases, glycosyltransferases, and other enzymes, to form various protopanaxadiol-type, protopanaxatriol-type, and oleanane-type ginsenosides. Like other saponins, it is believed that the oligosaccharide chains were likely to be synthesized by the sequential addition of single sugar molecules to the aglycone [82, 83]. Compared with that of protopanaxadiol-type ginsenosides, the aglycone of protopanaxatriol-type ginsenosides (protopanaxatriol) has one more hydroxyl group at C-6, which possibly stems

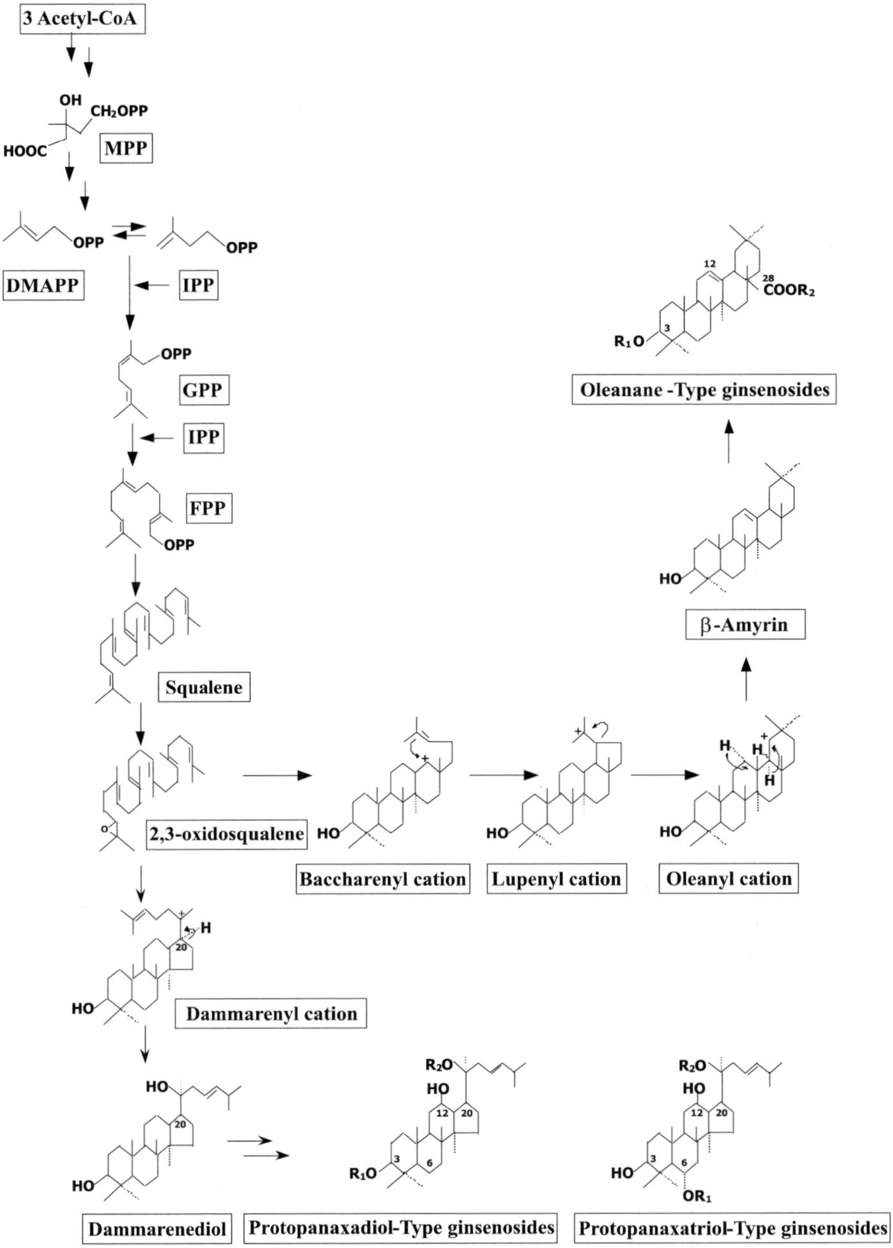

Fig. 7 The proposed ginsenoside biosynthetic pathway (modified from Refs. [80, 81])

from protopanaxadiol by oxidation. Glycosylation sites of protopanaxatriol are usually C-6 and C-20, but not C-3, at which glycosylation occurs for protopanaxadiol.

3.2.2 Manipulation of Ginsenoside Heterogeneity

Manipulation of ginsenoside heterogeneity has been performed in cell cultures, especially in *P. notoginseng* cell cultures. *P. notoginseng*, a famous traditional Chinese medicinal herb, is an important source of ginsenosides, and it has been used as a source of a healing drug and health tonic in oriental countries since ancient times. Ginsenosides, mostly protopanaxadiol-type and protopanaxatriol-type, are known as its major bioactive secondary metabolites. The main strategies for manipulation of individual ginsenoside biosynthesis are to intentionally change environmental factors in cell cultures.

3.2.2.1 Addition of Jasmonates

At present, the metabolic pathway engineering of ginseng cells for manipulation of the ginsenoside heterogeneity is very difficult, since it is not clear how each individual ginsenoside is synthesized. In a primary study, it was suggested that both the amount and the type of the ginsenoside produced by the cultured cells of *P. notoginseng* could be varied under different culture modes [124]. Elicitation of jasmonates proved to be an effective way to manipulate ginsenoside heterogeneity [84].

Different jasmonates play different roles in ginsenoside biosynthesis. Dihydromethyl jasmonate (HMJA) showed less effect than MJA on ginsenoside synthesis, and only the 100 μM concentration of HMJA increased the ginsenoside content. In contrast, MJA showed a significant effect, and more importantly, MJA changed the ratio of ginsenoside content. The content of ginsenoside Rb_1 increased much more than that of ginsenosides Rg_1 and Re did. In addition, Rd was easily detected upon the addition of MJA. The ratio of the Rb (protopanaxadiol-type) to the Rg (protopanaxatriol-type) groups of the ginsenosides increased from 0.67 (control) to 1.84 (at 100 μM MJA). In contrast, under HMJA elicitation, the ratio of Rb to Rg did not change significantly, and no Rd was detected. The results suggest that MJA is a promising compound for the manipulation of the heterogeneity of ginsenosides in *P. notoginseng* cell cultures [84].

The MJA concentration was also significant for the ginsenoside synthesis [84]. Table 5 presents the contents of different ginsenosides at MJA concentrations of 20–500 μM. MJA remarkably enhanced the ginsenoside content and altered its distribution in the cell cultures. The total ginsenoside content increased with increasing MJA concentration from 20 to 200 μM, then a slight decrease was observed at even higher concentrations of MJA. Upon addition of MJA, the ginsenoside content of the Rb group increased much more than that of the Rg group. In particular, the content of Rb_1 increased far more than

Table 5 Effects of methyl jasmonate (*MJA*) concentration on the production and distribution of individual ginsenosides

MJA concentration (μM)	Ginsenoside production (mg/L)					Rb:Rg[b]
	Rg_1	Re	Rb_1	Rd	Total[a]	
Day 12						
0	39.2 ± 1.4	34.0 ± 2.6	28.3 ± 1.9	0 ± 0	101 ± 6	0.39
0[c]	29.5 ± 1.1	29.9 ± 3.0	26.7 ± 2.4	0 ± 0	86.1 ± 6.5	0.45
20	68.0 ± 3.7	54.6 ± 1.4	114 ± 8	13.4 ± 3.3	250 ± 16	1.04
100	68.9 ± 3.6	54.6 ± 1.4	190 ± 18	23.5 ± 0.5	337 ± 24	1.72
200	68.7 ± 1.5	53.3 ± 2.2	226 ± 15	22.2 ± 5.9	370 ± 25	2.03
500	39.1 ± 0.4	26.8 ± 0.4	136 ± 10	5.99 ± 0.64	207 ± 11	2.07
Day 15						
0	25.1 ± 1.7	34.3 ± 2.3	29.1 ± 1.6	0 ± 0	88.5 ± 5.6	0.49
0[c]	27.9 ± 2.0	33.7 ± 0.8	38.3 ± 2.6	0 ± 0	99.9 ± 5.4	0.62
20	65.4 ± 11.0	61.4 ± 8.8	132 ± 16	9.12 ± 0.45	268 ± 36	1.12
100	65.9 ± 0.4	60.5 ± 0.5	195 ± 3	12.7 ± 1.2	333 ± 5	1.64
200	66.8 ± 0.0	64.7 ± 0.6	256 ± 6	15.9 ± 0.6	403 ± 7	2.06
500	36.7 ± 2.0	35.9 ± 2.1	164 ± 5	7.28 ± 0.39	244 ± 10	2.49

[a] Total content=(Rg1+Re+Rb1+Rd)
[b] Rb:Rg=(Rb1+Rd)/(Rg1+Re)
[c] The control with addition of 1mL/L ethanol, which was used for dissolving MJA

that of Rg_1 and Re, and Rd was also detected in all cases of MJA supplementation. An increase in MJA concentration from 0 to 500 μM resulted in an increase in the ratio of Rb to Rg from 0.39 to 2.07 on day 12 and from 0.49 to 2.49 on day 15. It was also observed that the ratio of Rb to Rg increased sharply with addition of 200 μM MJA, while there was no significant change for the control during the entire cultivation period (Fig. 8). The improvement of ginsenoside production and the alteration of ginsenoside distribution (heterogeneity) by jasmonate elicitation were also observed in adventitious root cultures of *P. ginseng* [85]. All those facts suggest that jasmonate as a signal transducer may activate major enzymes in the isoprenoid pathway up to dammarenediol and may also enhance key enzyme activities in the biosynthetic steps from dammarenediol to individual ginsenosides (especially Rb_1 and Rd).

The combination of MJA re-elicitation with sucrose feeding was demonstrated to be a simple and effective strategy for hyperproduction of ginsenosides and efficient manipulation of their heterogeneity in a bioreactor. The maximum cell dry weight (DW), the ginsenoside content when the cells reached their maximum DW, and the maximum ginsenoside production for the control, for MJA elicited twice and, for the combination strategy are summarized in Table 6. The maximum DW for the combination strategy was 25.1 ± 0.3 and 27.3 ± 1.5 g/L on day 17 in a flask and an airlift bioreactor (ALR), respectively, which was about 20 and 30% higher than for the control and for MJA elicited twice in both cases. Similar to MJA re-elicitation, in both cultivation vessels, the ginsenoside content was also highly enhanced with the combination strategy, and therefore higher ginsenoside production was obtained. For example, in the ALR with the combination strategy, the production of ginsenosides Rg_1, Re, Rb_1, and Rd was 118.4 ± 4.7, 117.2 ± 4.6, 290.2 ± 5.1, and 32.7 ± 8.1 mg/L, respectively, which was apparently higher

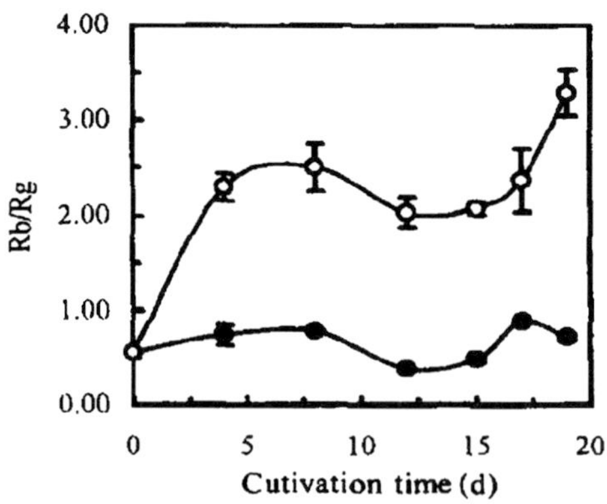

Fig. 8 Dynamic profiles of the ginsenoside Rb-to-Rg ratio in *Panax notoginseng* cell cultures. Control (*closed symbols*), methyl jasmonate (*MJA*) addition (*open symbols*)

Table 6 Effects of combination strategy on maximum dry weight (*DW*), individual ginsenoside content, and maximum production of individual ginsenosides

Cultivation conditions	Maximum DW (g/L)	Ginsenoside content (mg/100 mg DW) Rg_1	Re	Rb_1	Rd	Total[1]	Ginsenoside production (mg/L) Rg_1	Re	Rb_1	Rd
Flasks										
Control (day 15)	20.8 ± 0.8^{a}	0.24 ± 0.01^{a}	0.25 ± 0.02^{a}	0.24 ± 0.02^{a}	0^{a}	0.74 ± 0.03^{a}	50.3 ± 3.7^{a}	52.4 ± 1.0^{a}	50.9 ± 3.3^{a}	0^{a}
MJA elicited twice[2] (day 17)	18.9 ± 0.5^{b}	0.42 ± 0.01^{b}	$0.45 \pm 0.01^{b,c}$	1.17 ± 0.04^{b}	$0.11 \pm 0.03^{b,c}$	$2.15 \pm 0.07^{b,c}$	79.3 ± 4.8^{b}	85.0 ± 5.0^{b}	220.4 ± 2.2^{b}	20.8 ± 5.9^{b}
Combination strategy[3] (d 17)	25.1 ± 0.3^{c}	0.45 ± 0.01^{c}	0.46 ± 0.02^{b}	1.22 ± 0.03^{b}	0.14 ± 0.04^{b}	2.27 ± 0.05^{b}	112.9 ± 2.1^{c}	120.4 ± 2.9^{c}	306.1 ± 4.5^{c}	35.1 ± 6.9^{c}
ALR										
Control (day 15)	23.1 ± 1.6^{d}	0.21 ± 0.02^{d}	0.22 ± 0.01^{a}	0.22 ± 0.03^{a}	0^{a}	0.64 ± 0.05^{a}	48.5 ± 3.1^{a}	49.9 ± 3.4^{a}	49.8 ± 2.4^{a}	0^{a}
MJA elicited twice[2] (day 17)	21.3 ± 0.9^{a}	0.39 ± 0.02^{e}	0.42 ± 0.02^{c}	0.98 ± 0.04^{c}	0.09 ± 0.01^{c}	1.87 ± 0.10^{d}	82.1 ± 8.1^{b}	88.5 ± 8.3^{b}	209.0 ± 8.0^{b}	19.2 ± 3.8^{b}
Combination strategy[3] (day 17)	27.3 ± 1.5^{e}	$0.41 \pm 0.02^{b,e}$	$0.43 \pm 0.01^{b,c}$	1.06 ± 0.07^{d}	$0.12 \pm 0.04^{b,c}$	$2.02 \pm 0.06^{c,d}$	111.8 ± 4.7^{c}	117.2 ± 4.6^{c}	290.2 ± 5.1^{c}	32.7 ± 8.1^{c}

a, *b*, *c*, *d*, and *e* means with the same *letter* all noted in a single column are not significantly different according to Tukey's honestly significant difference multiple-comparison test with a family error rate of 0.05.

[1]Total content = (Rg_1+Re+Rb_1+Rd)

[2]MJA re-elicitation: 200 μM of MJA added on days 8 and 13, respectively

[3]Combination strategy: 200 μM of MJA added on days 8 and 13 with feeding of 10 g sucrose/L on day 13

than for the control and for MJA re-elicitation. The results show that MJA re-elicitation combined with sucrose feeding was also suitable for the bioreactor cultivation of *P. notoginseng* cells for hyperproduction of heterogeneous ginsenosides [86].

Furthermore, our laboratory has used novel chemically synthesized 2-hydroxyethyl jasmonate (HEJA) to induce the ginsenoside biosynthesis and to manipulate the product heterogeneity in cell suspension cultures of *P. notoginseng* [87]. It was interestingly found that HEJA could stimulate ginsenoside biosynthesis and change the heterogeneity more efficiently than MJA, and the activity of the Rb_1 biosynthetic enzyme, i.e., UDPG:ginsenoside Rd glucosyltransferase (UGRdGT), was also higher in the former case (Fig. 9). By investigating two signal events in the plant defense response, i.e., oxidative burst and jasmonic acid (JA) biosynthesis, the results suggest that an oxidative burst might not be involved in the jasmonate-elicited signal transduction pathway, and MJA and HEJA may induce the ginsenoside biosynthesis via induction of endogenous JA biosynthesis and key enzymes in the ginsenoside biosynthetic pathway such as UGRdGT. The information is considered useful for hyperproduction of plant-specific heterogeneous products.

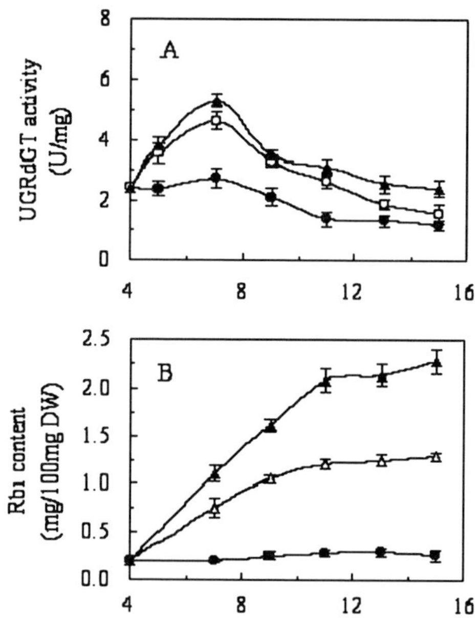

Fig. 9 a Dynamic changes of UDPG:ginsenoside Rd glucosyltransferase (*UGRdGT*) activity **b** and the content of ginsenoside Rb_1 for *P. notoginseng* cells with 200 μM MJA or 2-hydroxyethyl jasmonate (*HEJA*) elicited on day 4. Control (*circles*), 200 μM MJA added on day 4 (*open triangles*), 200 μM HEJA added on day 4 (*closed triangles*)

3.2.2.2
Change of Oxygen Partial Pressure

Although the oxygen requirement of plant cells is relatively modest compared with that of microbial cells, high cell density and fluid viscosity could significantly reduce the oxygen transfer efficiency in bioreactors. An alternative approach to avoid oxygen limitation in bioreactors is via manipulation of oxygen partial pressure (pO_2). Different pO_2 levels could be obtained by mixing air with different ratios of pure oxygen or nitrogen while the total aeration rate was maintained constant. Different pO_2 levels affected the distribution of ginsenosides (heterogeneity) in high-density cell cultures in 1-L ALRs [88]. On day 10, the ratio of Rb to Rg at pO_2 of 36.5 kPa is 1.8- and 1.5-fold that at pO_2 of 10.6 and 21.3 kPa, respectively, while supplementation of CO_2 at pO_2 of 10.6 and 36.5 kPa had no obvious effects on ginsenoside formation. The results imply that pO_2 may play an interesting role in ginsenoside biosynthesis via signal transduction like an oxidative burst [88].

3.2.2.3
Change of External Calcium Concentration

Calcium is considered as the most versatile intracellular messenger, and is able to couple a wide range of extracellular signals to specific responses [89]. In recent years, evidence has suggested that extracellular Ca^{2+} affects plant secondary metabolite production [90, 91]. It was observed that external calcium not only affected biosynthesis of ginsenoside Rb_1 [92], but also changed the Rb to Rg ratio (Table 7). External calcium affected the content of intracellular calcium and calmodulin (CaM) and the activities of calcium-dependent protein kinases (CDPKs) and key enzymes leading to ginsenoside heterogeneity, e.g., ginsenoside glycosyltransferases such as UGRdGT [92]. It is proposed that the effects of external calcium on the ginsenoside biosynthesis by *P. notoginseng* cells are possibly mediated via a signal transduction pathway (Fig. 10). Regulation of the external calcium concentration is considered as a useful and powerful tool for manipulating ginsenoside synthesis and its heterogeneity in a large-scale cultivation process.

3.2.2.4
Biotransformation

The distribution of various ginsenosides in ginseng cells is very different, and unfortunately the rare ginsenosides usually present higher physiological activity than the abundant ones. For example, ginsenoside Rh_2, whose content in wild ginseng is around 0.00003 (by dry weight), shows stronger potency to inhibit tumor growth than that of ginsenoside Rb_1, whose content is around 0.01. To date, it is very difficult to manipulate the accumulation of rare gin-

Table 7 Effects of external calcium concentration on the distribution of individual ginsenosides

Initial Ca^{2+} concentration (mM)	Verapamil addition or Ca^{2+} feeding	Rb:Rg[a] 0 h	24 h	48 h	72 h
0	–	0.43	0.42	0.43	0.43
3	–	0.43	0.45	0.49	0.51
8	–	0.43	0.48	0.57	0.61
13	–	0.43	0.44	0.45	0.48
3	Addition of 0.5 mM Verapamil at initial time	0.43	0.42	0.43	0.47
3	Feeding of 5 mM Ca^{2+} at 24 h	0.43	0.42	0.57	0.66
3	Feeding of 5 mM Ca^{2+} at 24 and 48 h	0.43	0.42	0.57	0.57

[a] Rb:Rg=Rb1/(Rg1+Re)

senosides in ginseng cells as their biosynthetic process is unclear. Biotransformation is a practical approach to transform highly abundant ginsenosides into rare ones by using isolated enzymes or microorganisms. Table 8 shows

Table 8 Biotransformation of ginsenosides by enzymes or microorganisms

Transformation of ginsenosides	Enzymes or microorganisms
$Rg_3 \rightarrow Rh_2$	Ginsenoside-β-glucosidase (from *Panax ginseng*) [93] *Rhizopus stolonifer* AS 3.822 [94] *Bacteroides* sp., *Fusobacterium* sp., *Bifidobacterium* sp. [95]
$Rc \rightarrow Rd$	Ginsenoside-α-arabinofuranase (from *P. ginseng*) [96]
$Rg_2 \rightarrow Rh_1$	Ginsenoside-α-L-rhamnosidase (from *Absidia* sp.39) [97]
$Rb_1 \rightarrow F_2$	Ginsenoside-β-glucosidase (from *Fusobacterium* K-60) [98]
$Rg_1, Re \rightarrow Rh_1$	Lactase (from *Penicillium sp.*) [99]
$Rb_2 \rightarrow Rd$	α-L-Arabinopyranosidase (from *Bifidobacterium breve* K-110) [100]
$Rc \rightarrow Rd$	α-L-Arabinofuranosidase (from *B. breve* K-110) [100]
$Re \rightarrow Rg_1$	Hesperidinase (from *Penicillium* sp.) [101]
$Rb_1 \rightarrow Rd$	*Curvularia lunata* AS 3.4381, *R. stolonifer* AS 3.822 [94]
$Rd \rightarrow Rg_3$	*R. stolonifer* AS 3.822 [94]

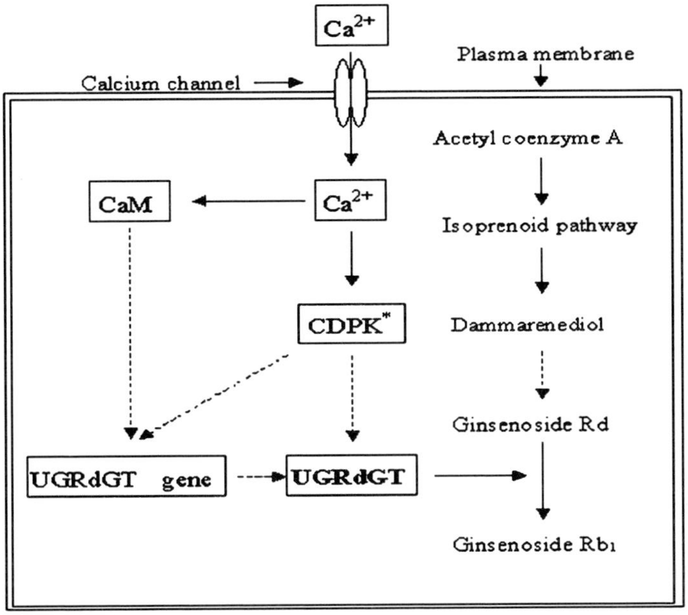

Fig. 10 A proposed signal transduction pathway regarding the effect of external Ca^{2+} on biosynthesis of ginsenoside Rb_1 by *P. notoginseng* cells. Ca^{2+} signal changes are triggered by various concentrations of external Ca^{2+}. The calcium signatures are decoded by calcium sensors, calmodulin (*CaM*) and calcium-dependent protein kinase (*CDPK*). UGRdGT, which catalyzes ginsenoside Rb_1 synthesis from Rd, is possibly modulated by the sensors in a direct or an indirect way (*dashed lines*). Changes of CDPK activity may result from increased synthesis of CDPK protein or from post-translational modification of the enzyme (*CDPK**)

some enzymes and microorganisms used in ginsenoside biotransformation. High biotransformation rates have been observed. For example, after reaction at 60 °C for 24 h, over 60% of ginsenoside Rg_3 was converted to Rh_2 by ginsenoside-β-glucosidase from ginseng [93]. After 4-day incubation on a rotary shaker (200 rpm) at 24 °C with *Curvularia lunata*, 81% of ginsenoside Rb_1 was transformed into Rd [94]. Besides hydrolyzing the ginsenosides conjugated with many sugars to that conjugated with fewer sugars, glycosylation on the ginsenosides with a few sugars is another method of ginsenoside biotransformation. The UGRdGT isolated from *P. notoginseng* cell cultures in our laboratory allowed over 80% of ginsenoside Rd to produce Rb_1 after reaction at 30 °C for 10 h with uridine 5′-diphosphoglucose. Although both isolated enzymes and microorganisms can convert ginsenosides, the products of ginsenoside biotransformation by enzymes are single ones and its incubation time is also shorter than for conversion by microorganisms. Thus, the biotransformation by enzymes is a promising approach in the manipulation of ginsenoside heterogeneity. But, its disadvantage is that another ginsenoside (as a substrate) and the enzyme (as a biocatalyst) are necessary, which may cause a high cost especially for large-scale production.

4
Perspectives

As we gain deeper insight into the metabolic network and its interaction with the environment of biosynthetic pathways for plant secondary metabolism, more rational approaches to redirecting metabolic flux to desired secondary metabolites could be designed. By integrating molecular biology techniques with mathematical analysis tools, we can use metabolic engineering to help elucidate metabolic flux control and rational selection of targets for genetic modification [102, 103]. In the case of plant alkaloids (one of the largest groups of natural products), which provide many pharmacologically active compounds, significant progress, such as increased indole alkaloid levels, altered tropane alkaloid accumulation, elevated serotonin synthesis, reduced indole glucosinolate production, redirected shikimate metabolism, and increased cell-wall-bound tyramine formation, has been achieved by metabolic engineering applications [104–107].

Functional genomics (transcriptomics, proteomics, and metabolomics) also offer new avenues for potential manipulation of heterogeneity of plant secondary metabolites. Because not enough genomic tools are available for most plants producing interesting secondary metabolites (e.g., ginsenosides and paclitaxel), despite great progress in cDNA cloning of enzymes related to biosynthesis of paclitaxel [108], it is not surprising that virtually no such comprehensive studies have been reported. Recently, a proteomic approach was taken to analyze the proteins in opium poppy latex, which is thought to be the major site of morphine biosynthesis [109]. This type of analysis based on two-dimensional sodium dodecyl sulfate–polyacrylamide gel electrophoresis is helpful to identify the genes required for specific cell factories that are responsible for the biosynthesis of plant secondary metabolites such as morphine. It is very important to analyze the protein itself closely related to secondary metabolism, because the DNA sequence and the expression of messenger RNA (mRNA) do not provide information of protein post-translational modification, structure, and protein–protein interaction. Almost all proteins are post-translationally modified, and then form specific structures and functions through protein–protein interaction [110]. In addition, transcriptomics tools such as differential display, expressed sequence tag databases and microarrays have also been used to investigate the biosynthesis of specific secondary metabolites, and, in particular, random sequencing of cell cDNA libraries from MJA-induced *T. cuspidata* cells for taxoid biosynthesis has been used to isolate the entire paclitaxel pathway [108, 111–113].

Considering the network of the biosynthetic pathway of plant secondary metabolites, the same metabolite can be a member of several different pathways and may also have regulatory effects on multiple biological processes. Therefore, an individual metabolite cannot, in most cases, be unambiguously linked to a single genomic sequence [114]. Thus, the simultaneous identification and quantification of metabolites is necessary to study the dynamics of the metabolome of secondary metabolism, to analyze fluxes in secondary metabolic pathways, and to decipher the role of each metabolite following various stimuli. Linkage of functional metabolomic information to mRNA and protein expression data makes it possible to visualize the functional genomic repertoire of cells [115]. Such knowledge is believed to have great potential for manipulation of heterogeneity of plant secondary metabolites.

In the postgenomic era, the processes and strategies to manipulate plant cell cultures for heavy accumulation of desired secondary metabolites such as Tc are possibly like the following: establishment of cell cultures able to produce Tc; determination of suitable cultivation conditions, for example, elicitation with novel synthetic jasmonates [116, 117] or other stimuli which activate the genes involved in Tc biosynthesis and enhance Tc production; metabolite profiling by means of gas chromatography–mass spectrometry (MS), liquid chromatography–MS, NMR, and so on; proteomic analysis; discovery of genes related to Tc accumulation by means of cDNA–amplified fragment length polymorphism, serial analysis of gene expression and microarrays, and integration with proteome analysis data; enhancement of expression or activity of rate-limiting enzymes via transformation with selected genes alone or in combination; decrease of the flux through competitive pathways and the catabolism of Tc and prevention of feedback inhibition of a key enzyme via manipulation by transcription factors or antisense technology; and combination with engineering strategies such as pulsed electric field stimulation [118].

Until now, only a few of the these strategies have been successfully demonstrated in plant cells. Recently, the simultaneous overexpression of two genes encoding the rate-limiting upstream enzyme putrescine *N*-methyltransferase and the hyoscyamine-6β-hydroxylase of tropane alkaloid biosynthesis resulted in the highest scopolamine production ever obtained in cultivated *H. niger* hairy roots [119]. Antisense approaches and transcription factors were also successfully applied to manipulation of secondary metabolite production [120, 121]. Because transcription factors are efficient new molecular tools for plant metabolic engineering to increase the production of valuable compounds, the use of specific transcription factors would avoid the time-consuming step of acquiring knowledge about all enzymatic steps of a poorly characterized biosynthetic pathway [122]. For example, high-flavonol tomatoes were obtained via the heterologous expression of the maize transcription factor genes [123]. It is expected that very efficient production of high-value-added secondary metabolites by plant cells will be possible with the advancement of functional genomic technology.

Acknowledgements W. Wang contributed to our ginsenoside heterogeneity project. Financial support from the National Natural Science Foundation of China (NSFC project nos. 30270038 and 20236040) and the Shanghai Science & Technology Commission (project no. 04QMH1410) is gratefully acknowledged. J.J.Z. also thanks the National Science Fund for Distinguished Young Scholars (NSFC project no. 20225619) and the Cheung Kong Scholars Program of the Ministry of Education of China.

References

1. Hostettmann K, Terreaux C (2000) Search for new lead compounds from higher plants. Chimia (Aarau) 54:652–657
2. Verpoorte R (1998) Exploration of nature's chemodiversity: the role of secondary metabolites as leads in drug development. Drug Discov Today 3:232–238
3. De Luca V, St Pierre B (2000) The cell and developmental biology of alkaloid biosynthesis. Trends Plant Sci 5:168–173
4. Wink M (1998) Plant breeding: importance of plant secondary metabolites for protection against pathogens and herbivores. Theor Appl Genet 75:225–233
5. Harborne JB, Baxter H (1999) The handbook of natural flavonoids, vol 1. Wiley, Chichester
6. Buckingham J (ed) (2000) Dictionary of natural products on CD. Chapman & Hall/CRC, UK
7. Ibrahim RK, Varin L (1993) Flavonoid enzymology. In: Lea PJ (ed) Methods in plant biochemistry, vol 9. Academic, London, pp 99–131
8. Facchini PJ (1999) Plant secondary metabolism: out of the evolutionary abyss. Trends Plant Sci 4:382–384
9. Osbourne AE, Wubben PJ, Melton RE, Carter JP, Daniels MJ (1998) Saponins and plant defense. In: Romeo TJ, Downum KR, Verpoorte R (eds) Phytochemical signal and plant-microbe interactions. Plenum, New York, pp 1–16
10. Chappell J (1995) Biochemistry and molecular biology of the isoprenoid biosynthetic pathway in plants. Annu Rev Plant Physiol Plant Mol Biol 46:521–547
11. Croteau R, Kutchan TM, Lewis NG (2000) Natural products (secondary metabolites). In: Buchanan B, Gruissem W, Jones R (eds) Biochemistry and molecular biology of plants. ASPB, Rockville, MD, pp 1250–1268
12. McGarvey DJ, Croteau R (1995) Terpenoid metabolism. Plant Cell 7:1015–1026
13. Kingston DGI (2001) Taxol, a molecule for all seasons. Chem Commun 867–880
14. Zheng GZ, Yang CFL (1994) Sanchi (Punux notoginseng): biology and application. Science, Beijing (in Chinese)
15. Sticher O (1998) Getting to the root of ginseng. CHEMTECH 28:26–32
16. Stafford AM, Pazoles CJ, Siegel S, Yeh L-A (1998) Plant cell culture: a vehicle for drug discovery. In: Harvey AL (ed) Advances in drug techniques. Wiley, New York, pp 53–64
17. Wani MC, Taylor HL, Wall ME, Coggon P, McPhail AT (1971) Plant antitumour agents VI. The isolation and structure of taxol, a novel antileukemic and antitumour agent from Taxus brevifolia. J Am Chem Soc 93:2325–2327

18. Miller RW, Powell RG, Smith CR, Arnold E, Clardy J (1981) Antileukemic alkaloids from Taxus wallichiana Zucc. J Org Chem 46:1469–1474
19. Witherup KM, Look SA, Stasko MW, Ghiorzi TJ, Muschik GM (1990) Taxus spp.: needles contain amounts of taxol comparable to the bark of Taxus brevifolia: analysis and isolation. J Nat Prod 53:1249–1255
20. Fett-Neto AG, DiCosmo F (1992) Distribution and amount of taxol in different shoot parts of Taxus cuspidata. Planta Med 58:464–466
21. ElSohly HN, Croom ED, Kopycki WJ, Joshi AS, ElSohly MA, McChesney JD (1995) Concentrations of taxol and related taxanes in the needles of different Taxus cultivars. Phytochem Anal 6:149–156
22. Singh B, Gujral RK, Sood RP, Duddeck H (1997) Constituents from Taxus species. Planta Med 63:191–192
23. Strobel GA, Ford E, Li JY, Sears J, Sidhu RS, Hess WM (1999) Seimatoantlerium tepuiense gen. nov., a unique epiphytic fungus producing taxol from the Venezuelan-Guayana system. Appl Microbiol 22:426–433
24. Wang J, Li G, Lu H, Zheng Z, Huang Y, Su W (2000) Taxol from Tubercularia sp. strain 333 TF5, an endophytic fungus of Taxus mairei. FEMS Microbiol Lett 193:249–253
25. Shrestha K, Strobel GA, Prakash S, Gewali M (2001) Evidence for paclitaxel from three new endophytic fungi of Himalayan yew of Nepal. Planta Med 6 7:374–376
26. Baloglu E, Kingston DGI (1999) The taxane diterpenoids. J Nat Prod 62:1448–1472
27. Sledge GW (2003) Gemcitabine combined with paclitaxel or paclitaxel/trastuzumab in metastatic breast cancer. Semin Oncol 30(Suppl):19–21
28. O'Brien MER, Splinter T, Smit EF, Biesma B, Krzakowski M, Tjan-Heijnen VCG, Van Bochove A, Stigt J, Smid-Geirnaerdt MJA, Debruyne C, Legrand C, Giaccone G (2003) Carboplatin and paclitaxol (Taxol) as an induction regimen for patients with biopsy-proven stage IIIA N2 non-small cell lung cancer: an EORTC phase II study (EORTC 08958). Eur J Cancer 39:1416–1422
29. Guéritte F (2001) General and recent aspects of the chemistry and structure-activity relationships of taxoids. Curr Pharm Design 7:1229–1249
30. Schiff PB, Fant J, Horwitz SB (1979) Promotion of microtubule assembly invitro by taxol. Nature 277 (5698):665–667
31. Kingston DGI (2000) Recent advances in the chemistry of taxol. J Nat Prod 63:726–734
32. Shigemori H, Kobayashi J (2004) Biological activity and chemistry of taxoids from the Japanese yew, Taxus cuspidate. J Nat Prod 67:245–256
33. Eisenreich W, Menhard B, Hylands PJ, Zenk MH, Bacher A (1996) Studies on the biosynthesis of taxol: the taxane carbon skeleton is not of mevalonoid origin. Proc Natl Acad Sci USA 93:6431–6436
34. Eisenreich W, Rohdich F, Bacher A (2001) Deoxyxylulose phosphate pathway to terpenoids. Trends Plant Sci 6:78–84
35. Rohmer M, Knani M, Simonin P, Sutter B, Sahm H (1993) Isoprenoid biosynthesis in bacteria: a novel pathway for the early steps leading to isopentenyl diphosphate. Biochem J 295:517–524
36. Lichtenthaler HK, Rohmer M, Schwender J (1997) Two independent biochemical pathways for isopentenyl diphosphate and isoprenoid biosynthesis in higher plants. Physiol Plant 101:643–652
37. Lichtenthaler HK (1999) The 1-deoxy-D-xylulose-5-phosphate pathway of isoprenoid biosynthesis in plants. Annu Rev Plant Physiol Plant Mol Biol 50: 47-65

38. Koepp AE, Hezari M, Zajicek J, Stofer-Vogel B, LaFever RE, Lewis NG, Croteau R (1995) Cyclization of geranylgeranyl diphosphate to taxa-4(5),11(12)-diene is the committed step of taxol biosynthesis in Pacific yew. J Biol Chem 270:8686–8690
39. Hezari M, Lewis NG, Croteau R (1995) Purification and characterization of taxa-4(5),11(12)-diene synthase from Pacific yew (Taxus brevifolia) that catalyses the first committed step of Taxol biosynthesis. Arch Biochem Biophys 322:437–444
40. Hezari M, Ketchum REB, Gibson DM, Croteau R (1997) Taxol production and taxadiene synthase activity in Taxus canadensis cell suspension cultures. Arch Biochem Biophys 337:185–190
41. Dong HD, Zhong JJ (2001) Significant improvement of taxane production in suspension cultures of Taxus chinensis by combining elicitation with sucrose feed. Biochem Eng J 8:145–150
42. Hefner J, Rubenstein SM, Ketchum REB, Gibson DM, Williams RM, Croteau R (1996) Cytochrome P450-catalyzed hydroxylation of taxa-4(5),11(12)-diene to taxa-4(20),11(12)-diene-5α-ol: the first oxygenation step in taxol biosynthesis. Chem Biol 3:479–488
43. Jennewein S, Rithner CD, Williams RM, Croteau RB (2001) Taxol biosynthesis: Taxane 13α-hydroxylase is a cytochrome P450-dependent monooxygenase. Proc Natl Acad Sci USA 98:13595–13600
44. Walker KD, Ketchum REB, Hezari M, Gatfield D, Goleniowski M, Barthol A, Croteau R (1999) Partial purification and characterization of acetyl coenzyme A: taxa-4(20),11(12)-dien-5α-ol-*o*-acetyl-transferase that catalyses the first acetylation step of taxol biosynthesis. Arch Biochem Biophys 464:273–279
45. Jennewein S, Rithner CD, Williams RM, Croteau R (2003) Taxoid metabolism: taxoid 14β-hydroxylase is a cyto-chrome P450-dependent monooxygenase. Arch Biochem Biophys 413:262–270
46. Chau M, Jennewein S, Walker K, Croteau R (2004) Taxol biosynthesis: molecular cloning and characterization of a cytochrome P450 taxoid 7β-hydroxylase. Chem Biol 11:663–672
47. Floss HG, Mocek U (1995) Biosynthesis of taxol. In: Suffness M (ed.) Taxol science and applications. CRC, Boca Raton, pp 191–298
48. Kingston DGI, Molinero AA, Rimoldi JM (1993) The taxane diterpenoids. Prog Chem Org Nat Prod 61:1–206
49. Della Casa De Marcano DP, Halsall TG (1970) Crystallographic structure determination of the diterpenoid baccatin-V, a naturally occurring oxetane with a taxane skeleton. Chem Commum 1382–1383
50. Guéritte-Voegelein F, Guénard D, Potier P (1987) Taxol and derivatives: a biogenetic hypothesis. J Nat Prod 50:9–18
51. Walker K, Long R, Croteau R (2002) The final acylation step in taxol biosynthesis: cloning of the taxoid C13-side-chain N-benzoyltransferase from Taxus. Proc Natl Acad Sci USA 99:9166–9171
52. Walker K, Croteau R (2001) Taxol biosynthetic genes. Phytochemistry 58:1–7
53. Chau M, Croteau R (2004) Molecular cloning and characterization of a cytochrome P450 taxoid 2a-hydroxylase involved in Taxol biosynthesis. Arch Biochem Biophy 427:48–57
54. McCaskill D, Croteau R (1999) Isopentenyl diphosphate is the terminal product of the deoxyxylulose-5-phosphate pathway for terpenoid biosynthesis in plants. Tetrahedron lett 40:653–656

55. Choi HK, Kim SI, Son JS, Hong SS, Lee HS, Lee HJ (2000) Enhancement of paclitaxel production by temperature shift in suspension culture of Taxus chinensis. Enzyme Microb Technol 27:593–598
56. Bai J, Kitabatake M, Toyoizumi K, Fu L, Zhang S, Dai J, Sakai J, Hirose K, Yamori T, Tomida A, Tsuruo T, Ando M (2004) Production of biologically active taxoids by a callus culture of Taxus cuspidate. J Nat Prod 67:58–63
57. Ketchum REB, Rithnerb CD, Qiua D, Kima YS, Williamsb RM, Croteaua RB (2003) Taxus metabolomics: methyl jasmonate preferentially induces production of taxoids oxygenated at C-13 in Taxus x media cell cultures. Phytochemistry 62:901–909
58. Ketchum REB, Gibson DM, Croteau RB, Shuler ML (1999) The kinetics of taxoid accumulation in cell suspension cultures of Taxus following elicitation with methyl jasmonate. Biotech Bioeng 62:97–105
59. Veeresham C, Mamatha R, Prasad Babu Ch, Srisilam K, Kokate CK (2003) Production of taxol and its analogues from cell cultures of Taxus wallichiana. Pharm Biol 41:426–430
60. Brincat MC, Gibson DM, Shuler ML (2002) Alterations in taxol production in plant cell culture via manipulation of the phenylalanine ammonia lyase pathway. Biotechnol Prog 18:1149–1156
61. Dai JU, Cui J, Zhu WH, Guo HZ, Ye M, Hu Q, Zhang DY, Zheng JH, Guo D (2002) Biotransformation of 2α-, 5α-, 10β-, 14β-tetra-tetraacetoxy-4(20), 11-taxadiene by cell suspension cultures of Catharanthus roseus. Planta Med 68:1113–1117
62. Dai JG, Guo HZ, Ye M, Zhu WH, Zhang DY, Hu Q, Han J, Zheng JH, Guo DA (2003) Biotransformation of 4(20),11-taxadienes by cell suspension cultures of Platycodon grandiflorum. J Asian Nat Prod Res 5:5–10
63. Dai JG, Zhang SJ, Sakai J, Bai J, Oku Y, Ando M (2003) Specific oxidation of C-14 oxygenated 4(20), 11-taxadienes by microbial transformation. Tetrahedron Lett 44:1091–1094
64. Hu SH, Tian XF, Zhu WH, Fang QC (1996) Biotransformation of 2α-, 5α-, 10β-, 14β-tetra-tetraacetoxy-4(20), 11-taxadiene by the fungi Cunninghamella elegans and Cunninghamella echinulata. J Nat Prod 59:1006–1009
65. Hu SH, Tian XF, Zhu WH, Fang QC (1996) Microbial transformation of taxoids: Selective deacetylation and hydroxylation of 2α-, 5α-, 10β-, 14β-tetra-acetoxy-4(20),11-taxadiene by the fungus Cunninghamella echinulata Tetrahedron 52:8739–8746
66. Dai JG, Ye M, Guo HZ, Zhu WH, Zhang DO, Hu Q, Zheng JH, Guo D (2002) Regio- and stereo-selective biotransformation of 2α-,5α-,10β-, 14β-tetra-acetoxy-4(20), 11-taxadiene by Ginkgo cell suspension cultures. Tetrahedron 58:5659–5668
67. Hu SH, Tian XF, Zhu WH, Fang QC (1997) Biotransformation of some taxoids with oxygen substituent at C-14 by Cunninghamella echinulata. Biocatal Biotransform 14:241–250
68. Patel RN (1998) Tour de paclitaxel: Biocatalysis for semisynthesis. Annu Rev Microbiol 52:361–395
69. Patel RN, Banerjee A, Nanduri V (2000) Enzymatic acetylation of 10-deacetylbaccatin III to baccatin III by C-10 deacetylase from Nocardioides luteus SC 13913. Enzyme Microb Technol 27:371–375
70. Hanson RL, Kant J, Patel RN (2004) Conversion of 7-deoxy-10-deacetylbaccatin-III into 6-alpha-hydroxy-7-deoxy-10-deacetylbaccatin-III by Nocardioides luteus. Biotechnol Appl Biochem 39:209–214

71. Huang Q, Roessner CA, Croteau R, Scotta AI (2001) Engineering Escherichia coli for the synthesis of taxadiene, a key intermediate in the biosynthesis of Taxol. Bioorg Med Chem 9:2237–2242
72. Besumbes Ó, Sauret-Güeto S, Phillips MA, Imperial S, Rodriguez-Concepción M, Boronat A (2004) Metabolic engineering of isoprenoid biosynthesis in Arabidopsis for the production of taxadiene, the first committed precursor of Taxol. Biotechnol Bioeng 88:168–175
73. Soldati F, Sticher O (1980) HPLC separation and quantitative determination of ginsenosides from Panax ginseng, Panax quinquefolium and from ginseng drug preparations. Planta Med 39:348–357
74. Banthorpe DV (1994) Terpenoids. In: Mann J (ed) Natural products. Longman, Essex, UK, pp 331–339
75. Shibata S (2001) Preventing activities of ginseng saponins and some related triterpenoid compounds. J Korean Med Sci 16:S28–37
76. Odashima S, Ohta T, Kohno H, Matsuda T, Kitagawa I, Abe H, Arichi S (1985) Control of phenotypic expression of cultured B16 melanoma cells by plant glycosides. Cancer Res 45:2781–2784
77. Kim YS, Kim DS, Kim SI (1998) Ginsenoside Rh_2 and Rh_3 induce differentiation of HL-60 cells into granulocytes: Modulation of protein kinase C isoforms during differentiation by ginsenoside Rh_2. Int J Biochem Cell Biol 30:327–338
78. Islam MR, Mahdi JG, Bowen ID (1997) Pharmacological importance of stereochemical resolution of enantiomeric drugs. Drug Saf 17:149–165
79. Kudo K, Tachikawa E, Kashimoto T, Takahashi E (1998) Properties of ginseng saponin inhibition of catecholamine secretion in bovine adrenal chromaffin cells. Eur J Pharmacol 341:139–44
80. Haralampidis K, Trojanowska M Osbourn AE (2002) Biosynthesis of triterpenoid saponins in plants. Adv Biochem Eng Biotechnol 75:31–49
81. Kushiro T, Ohno Y, Shibuya M, Ebizuka Y (1997) In vitro conversion of 2,3-oxidosqualene into dammarenediol by Panax ginseng microsomes. Biol Pharm Bull 20:292–294.
82. Paczkowski C, Wojciechowski ZA (1994) Glucosylation and galactosylation of diosgenin and solasodine by soluble glycosyltransferase(s) from Solanum-melongena leaves. Phytochemistry 35:1429–1434
83. Wojciechowski ZA (1975) Biosynthesis of oleanolic acid glycosides by subcellular fraction of Calendular officinalis seedlings. Phytochemistry 14:1749–1753
84. Wang W, Zhong JJ (2002) Manipulation of ginsenoside heterogeneity in cell cultures of Panax notoginseng by addition of jasmonates. J Biosci Bioeng 93:48–53
85. Yu KW, Gao W, Hahn EJ, Paek KY (2002) Jasmonic acid improves ginsenoside accumulation in adventitious root culture of Panax ginseng C.A. Meyer. Biochem Eng J 11:211–215
86. Wang W, Zhang ZY, Zhong JJ (2005) Enhancement of ginsenoside biosynthesis in high density cultivation of Panax notoginseng cells by various strategies of methyl jasmonate elicitation. Appl Microbiol Biotechnol 67:752–758
87. Wang W (2004) Efficient induction of ginsenoside biosynthesis and manipulation of ginsenoside heterogeneity in cell suspension cultures of Panax notoginseng by addition of jasmonates. PhD thesis, ECUST, Shanghai
88. Han J, Zhong JJ (2003) Effects of oxygen partial pressure on cell growth and ginsenoside and polysaccharide production in high density cell cultures. Enzyme Microb Technol 32:498–503

89. Sanders D, Brownlee C, Harper JF (1999) Communicating with calcium. Plant Cell 11:691–706
90. Piñol MT, Palazón J, Cusidó RM, Ribó M (1999) Influence of calcium ion-concentration in the medium on tropane alkaloid accumulation in Datura stramonium hairy roots. Plant Sci 141:41–49
91. Nakao M, Ono K, Takio S (1999) The effect of calcium on flavanol production in cell suspension cultures of Polygonum hydropiper. Plant Cell Rep 18:759–776
92. Yue CJ, Zhong JJ (2005) Impact of external calcium and calcium sensors on ginsenoside Rb_1 biosynthesis by Panax notoginseng cells. Biotechnol Bioeng 89:444–452
93. Zhang C, Yu H, Bao Y, An L, Jin F (2001) Purification and characterization of ginsenoside-β-glucosidase from ginseng. Chem Pharm Bull 49:795–798
94. Dong A, Ye M, Guo H, Zheng H, Guo J (2003) Microbial transformation of ginsenoside Rb_1 by Rhizopus stolonifer and Curvularia lunata. Biotechnol Lett 25:339–344
95. Bae EA, Han MJ, Kim EJ, Kim DH (2004) Transformation of ginseng saponins to ginsenoside Rh_2 by acids and human intestinal bacteria and biological activities of their transformants. Arch Pharm Res 27:61–67
96. Zhang C, Yu H, Bao Y, An L, Jin F (2002) Purification and characterization of ginsenoside-α-arabinofuranase hydrolyzing ginsenoside Rc into Rd from the fresh root of Panax ginseng. Process Biochem 37:793–798
97. Yu H, Gong J, Zhang C, Jin F (2002) Purification and characterization of ginsenoside-α-L-rhamnosidase. Chem Pharm Bull 50:175–178
98. Park SY, Bae EA, Sung JH, Lee SK, Kim DH (2001) Purification and characterization of ginsenoside Rb_1-metabolizing β-glucosidase from Fusobacterium K-60, a human intestinal anaerobic bacterium. Biosci Biotechnol Biochem 65:1163–1169
99. Ko SR, Suzuki Y, Choi KJ, Kim YH (2000) Enzymatic preparation of genuine prosapogenini, 20(S)-ginsenoside Rh_1, from ginsenosides Re and Rg_1. Biosci Biotechnol Biochem 64:2739–2743
100. Shin HY, Park SY, Sung JH, Kim DH (2003) Purification and characterization of α-L-arabinopyranosidase and α-L-arabinofuranosidase from Bifidobacterium breve K-110, a human intestinal anaerobic bacterium metabolizing ginsenoside Rb_2 and Rc. Appl Environ Microbiol 69:7116–7123
101. Ko SR, Choi KJ, Uchida K, Suzuki Y (2003) Enzymatic preparation of ginsenosides Rg_2, Rh_1, and F_1 from protopanaxatriol-type ginseng saponin mixture. Planta Med 69:285–286
102. Stephanopoulos GN, Aristidou AA, Nielsen JE (1998) Metabolic engineering: principles and methodologies. Academic, New York
103. Nielsen J (ed) (2001) Metabolic engineering. Advances in Biochemical Engineering and Biotechnology, vol 73. Springer, Berlin Heidelberg New York
104. Yun DJ, Hashimoto T, Yamada Y (1992) Metabolic engineering of medicinal plants: transgenic Atropa belladonna with an improved alkaloid composition. Proc Natl Acad Sci USA 89:11799–11803
105. Sato F, Hashimoto T, Hachiya A, Tamura K, Choi KB, Morishige T, Fujimoto H, Yamada Y (2001) Metabolic engineering of plant alkaloid biosynthesis. Proc Natl Acad Sci USA 98:367–372
106. Facchini PJ (2001) Alkaloid biosynthesis in plants: biochemistry, cell biology, molecular regulation, and metabolic engineering applications. Annu Rev Plant Physiol Plant Mol Biol 52:29–66
107. Hughes EH, Hong SB, Gibson SI, Shanks JV, San KY (2004) Metabolic engineering of the indole pathway in Catharanthus roseus hairy roots and increased accumulation of tryptamine and serpentine. Metabol Eng 6:268–276

108. Jennewein S, Wildung MR, Chau M, Walker K, Croteau R (2004) Random sequencing of an induced Taxus cell cDNA library for identification of clones involved in Taxol biosynthesis. Proc Natl Acad Sci USA 101:9149–9154
109. Decker G, Wanner G, Zenk MH, Lottspeich F (2000) Characterization of proteins in latex of the opium poppy (Papaver somniferum) using two-dimensional gel electrophoresis and microsequencing. Electrophoresis 21:3500–3516
110. Hirano H, Islam, Kawasaki H (2004) Technical aspects of functional proteomics in plants. Phytochemistry 65:1487–1498
111. Yamazaki M, Saito K (2002) Differential display analysis of gene expression in plants. Cell Mol Life Sci 59:1246–1255
112. Suzuki H, Achnine L, Xu R, Matsuda SPT, Dixon RA (2002) A genomics approach to the early stages of triterpene saponin biosynthesis in Medicago truncatula. Plant J 32:1033–048
113. Guterman I, Shalit M, Menda N, Piestun D, Dafny-Yelin M, Shalev G, Bar E, Davydov O, Ovadis M, Emanuel M, Wang J, Adam Z, Pichersky E, Lewinsohn E, Zamir D, Vainstein A, Weiss D (2002) Rose scent: genomics approach to discovering novel floral fragrance-related genes. Plant Cell 14:2325–2338
114. Schwab W (2003) Metabolome diversity: too few genes, too many metabolites? Phytochemistry 62:837–849
115. Bino RJ, Hall RD, Fiehn O, Kopka J, Saito K, Draper J, Nikolau BJ, Mendes P, Roessner-Tunali U, Beale MH, Trethewey RN, Lange BM, Wurtele ES, Sumner LW (2004) Potential of metabolomics as a functional genomics tool. Trends Plant Sci 9:418–425
116. Qian ZG, Zhao ZJ, Tian WH, Xu Yf, Zhong JJ, Qian XH (2004) Novel synthetic jasmonates as highly efficient elicitors for taxoid production by suspension cultures of Taxus chinensis. Biotechnol Bioeng 86:595–599
117. Qian ZG, Zhao ZJ, Xu YF, Qian XH, Zhong JJ (2004) Novel chemically synthesized hydroxyl-containing jasmonates as powerful inducing signals for plant secondary metabolism. Biotechnol Bioeng 86:809–816
118. Ye H, Huang LL, Chen SD, Zhong JJ (2004) Pulsed electric field stimulates plant secondary metabolism in suspension cultures of Taxus chinensis Biotechnol Bioeng 88:788–795
119. Zhang L, Ding R, Chai Y, Bonfill M, Moyano E, Oksman-Caldentey KM, Xu T, Pi Y, Wang Z, Zhang H, Kai G, Liao Z, Sun X, Tang K (2004) Engineering tropane biosynthetic pathway in Hyoscyamus niger hairy root cultures. Proc Natl Acad Sci USA. 101:6786–6791
120. Chintapakorn Y, Hamill JD (2003) Antisense-mediated downregulation of putrescine N-methyltransferase activity in transgenic Nicotiana tabacum L. can lead to elevated levels of anatabine at the expense of nicotine. Plant Mol Biol 53:87–105
121. Van der Fits L, Memelink J (2000) ORCA3, a jasmonate responsive transcriptional regulator of plant primary and secondary metabolism. Science 289:295–297
122. Gantet P, Memelink J (2002) Transcription factors: tools to engineer the production of pharmacologically active plant metabolites. Trends Pharmacol Sci 23:563–569
123. Bovy A, de Vos R, Kemper M, Schijlen E, Pertejo MA, Muir S, Collins G, Robinson S, Verhoeyen M, Hughes S, Santos-Buelga C, van Tunen A (2002) High-flavonol tomatoes resulting from the heterologous expression of the maize transcription factor genes LC and C1. Plant Cell 14:2509–2526
124. Zhong JJ (1999) High-density cell cultivation and manipulation of heterogeneity of plant secondary metabolites. In: Proceedings of the APBioChEC, Phuket, Thailand, 1999

Adv Biochem Engin/Biotechnol (2005) 100: 89–179
DOI 10.1007/b136414

Published online: 5 July 2005

Model-based Inference of Gene Expression Dynamics from Sequence Information

Sabine Arnold[1] · Martin Siemann-Herzberg[2] · Joachim Schmid[2] · Matthias Reuss[2] (✉)

[1]Biotechnology R&D, DSM Nutritional Products Ltd., Bldg. 203/113A, 4002 Basel, Switzerland

[2]University of Stuttgart, Institute of Biochemical Engineering, Allmandring 31, 70569 Stuttgart, Germany
siemann@ibvt.uni-stuttgart.de, reuss@ibvt.uni-stuttgart.de

Abstract A dynamic model of prokaryotic gene expression is developed that makes considerable use of gene sequence information. The main contribution arises from the fact that the combined gene expression model allows us to access the impact of altering a nucleotide sequence on the dynamics of gene expression rates mechanistically. The high level of detail of the mathematical model is considered as an important step towards bringing together the tremendous amount of biological in-depth knowledge that has been accumulated at the molecular level, using a systems level analysis (in the sense of a bottom-up, inductive approach). This enables to the model to provide highly detailed insights into the various steps of the protein expression process and it allows us to access possible targets for model-based design. Taken as a whole, the mathematical gene expression model presented in this study provides a comprehensive framework for a thorough analysis of sequence-related effects on the stages of mRNA synthesis, mRNA degradation and ribosomal translation, as well as their nonlinear interconnectedness. Therefore, it may be useful in the rational design of recombinant bacterial protein synthesis systems, the modulation of enzyme activities in pathway design, in vitro protein biosynthesis, and RNA-based vaccination.

Keywords Dynamic modeling and simulation · Protein biosynthesis · Transcription · Translation · mRNA degradation

Abbreviations

Symbols

a_i	number of codons representing a particular amino acid *i*
A	number of naturally occurring amino acids
c	codon usage
C	metabolite concentration (μM)
d	spacing between ribosomes and degradosomes, and between SD sequence and translational start codons
D	promoter contained on DNA template
f	fraction of single-stranded bases within the 23 bases subsequent to the Shine-Dalgarno sequence
$f_{j,i}$	relative portion of base *j* contained in transcript *i* (%)
G	free energy (kJ/mol)
J	number of base triplets of a mRNA
k_i	respective rate constant
K	last codon of a coding region
K_{a}	association constant
K_{d}	dissociation constant
K_{I}	inhibition constant for respective metabolite (μM)
K_{M}	Michaelis-Menten constant for respective substrate (μM)
L_j	physical diameter of a ribosome and degradosome, respectively
m	mass (g)
m_i	ratio of RNA species *i* to total measured RNA (g/g)
$m_{i,j}$	element of matrix $\underline{\underline{M}}$
m_j	reference state of a ribosome and a degradosome, respectively
M	mRNA
M	number of mRNA molecules
$\underline{\underline{M}}$	mRNA matrix
n	number
n_i	transcript length for RNA species *i* (kb)
n_{cod}	number of base triplets used to denote a state
N	number of ribonucleic bases
N_{A}	Avogadro number
R	number of RNA species synthesized from a given DNA template

S	number of segments
t	time (min)
T	number of tRNA species
T	temperature (K)
T	time (s)
V	reaction rate (µM/min)
V	volume (µl)
V_P	relative protein expression rate (%)
X	measured radioactivity (dpm/µL)
z	position of endonucleolytic cleavage site
Z	number of fragments of a mRNA obtained by endonucleolytic cleavage

Greek letters

η	fractional codon usage
μ	specific growth rate (h^{-1})
Φ	efficiency factor
ϕ	T7 transcription terminator
$\phi 10$	T7 promoter
φ	energy charge

Indices

aq	aqueous
avg	average
cell	referring to a single cell
CR	catabolite repression
d	degradation
D	refers to promoter sequence of a DNA
D0	refers to a degradosome association site
dto	ditto
eff	effective
eq	thermodynamic equilibrium
exp	experimentally determined
f	formyl-
f	forward reaction
i	count index
in	entering equilibrium computation
I	induction
j	count index
k	count index
m	methionine
NTP	nucleoside triphosphate
out	outcome of equilibrium computation
qss	quasi-stationary state
r	reverse reaction
R0	refers to a ribosome binding site
s	count index
sim	predicted from simulation
t	denotes total concentration
un	unbound

Superscript

′	refers to new codon grid representation
0	initial condition
0	standard condition
A	refers to the A-site of a ribosome
D	degradosome
M	mRNA
M	methionine
max	maximum value
P	refers to the P-site of a ribosome
R	ribosome
R*	ribosome bound to the initiation codon prior to IF2-dissociation

Abbreviations

30S	small prokaryotic ribosomal subunit
30SIC	30S initiation complex
50S	large prokaryotic ribosomal subunit
70S	free, undissociated prokaryotic ribosome
70SIC	70S initiation complex
A	adenine
aa	amino acid(s)
aa-tRNA	aminoacyl-tRNA
Ac	acetate
Ack	acetate kinase
AcP	acetyl phosphate
ACSL	Advanced Continuous Simulation Language
Adk	adenylate kinase
ADP	adenosine diphosphate
Ala	alanine
AMP	adenosine monophosphate
Arg	arginine
ARS	aminoacyl-tRNA-synthetase
Asn	asparagine
Asp	aspartic acid
ass	association
ATP	adenosine triphosphate
AUG	translational start codon
bp	base pairs
BSA	bovine serum albumin
C	cytosine
CDP	cytosine diphosphate
CMP	cytosine monophosphate
CTP	cytosine triphosphate
Cys	cysteine
DNA	deoxyribonucleic acid
E	enzyme
EC	Enzyme Commission
EF	translational elongation factor
EMBL	European Molecular Biology Laboratory
endo	endonucleolytic

exo	exonucleolytic
F	folded conformation of the ribosome binding site
fMet-$tRNA_f^M$	*N*-formylmethionyl-tRNA
Frag	mRNA fragment
G	guanine
GDP	guanosine diphosphate
GFP	green fluorescent protein
Gln	glutamine
Glu	glutamic acid
Gly	glycine
GMP	guanosine monophosphate
GTP	guanosine triphosphate
h	hour
His	histidine
IC	initiation complex
IF	translational initiation factor
IF2D	IF2-dependent GTP hydrolysis
Ile	isoleucine
K	Kelvin
kb	kilobases
kDa	kiloDalton (1 Da $\hat{=}$ 1 g/mol)
kJ	kiloJoule
Leu	leucine
Lys	lysine
Met	methionine
min	minute
mRNA	messenger RNA
mv	degradosome movement
Ndk	nucleoside diphosphate kinase
NDP	nucleoside diphosphate
Nmk	nucleoside monophosphate kinase
NMP	nucleoside monophosphate
nt	nucleotide(s)
NTP	nucleoside triphosphate
P	promoter
PAGE	polyacryl amide gel electrophoresis
PAP I	poly-adenylate phosphorylase
pelB	pelB leader sequence
Phe	phenylalanine
Pi	inorganic phosphate
PNPase	polynucleotide phosphorylase
PPi	inorganic pyrophosphate
PPK	polyphosphate kinase
Pro	proline
RBS	ribosome binding site
rDNA	recombinant DNA
RF	translational termination factor
RFH	a particular translational termination factor
RNA	ribonucleic acid
RNAP	DNA-dependent RNA polymerase

RNase	ribonuclease
RP	ribosomal protein
RRF	ribosome release factor
rRNA	ribosomal RNA
s	second
S1	ribosomal protein S1 (contained in 30S ribosomal subunit)
Ser	serine
SNP	single-nucleotide polymorphism
ssRNA	single-stranded RNA
T	terminator
T	thymine
T	tRNA
T3	ternary complex (consists of one copy of EFTu, GTP, and aa-tRNA)
TC	transcription
TCA	tricarboxylic acid
TCE	transcription elongation
TCI	transcription initiation
TCT	transcription termination
TE	termination efficiency
THF	H_4-folate
Thr	threonine
TL	translation
TLE	translation elongation
TLI	translation initiation
TLT	translation termination
tmRNA	transfer-messenger RNA
Tris	tris(hydroxymethyl)aminomethane
tRNA	transfer RNA
Trp	tryptophan
Tyr	tyrosine
U	unit
U	uracil
UDP	uracil diphosphate
UMP	uracil monophosphate
UTP	uracil triphosphate
Val	valine

1 Introduction

The rapid advances in genomics research due to improved molecular biological, analytical and computational technologies have created a massive increase in the number of bioinformatic databases. Owing to the development of high-throughput DNA sequencing methods, complete genomes are now available for a variety of organisms. The primary reason for this tremendous interest and substantial progress is the fact that the genome of an entire organism contains, in its most condensed form, all the information necessary to construct this lifeform. It is the particular order of the nucleotides that comprise genomic DNA that specifies the uniqueness of an organism.

In the post-genomic era, great deal of the research in this area has been devoted to evaluate the functions of genes. Although efforts to systematically analyze these functions are underway, it has already been recognized that the analysis of these functions – and particularly the holistic functionality at the systems level – is much more complex than the genome sequencing itself was. However, tackling the most ambitious challenge in life science – to derive a relationship between the genome sequence information and nonlinear cellular dynamics – is even more complex. Understanding the link between genome sequence and protein expression levels is a first and essential prerequisite for a quantitative description of more complex phenomena. It should thus, in principle, be possible to derive the entire spectrum of cellular functionality and phenomena observed, including dynamic behavior, on the basis of genomic sequence information. At the same time, modeling and simulation of gene expression are also important in that they can be used to predict suitable strategies for genetic modification during the optimum design of expression systems.

The extent of protein expression is in many ways critically influenced by the encoded gene sequence. Regulatory elements at the initiation and termination sites of both the transcription and translation process are known to affect overall protein expression rate. However, the causes of differential mRNA degradation can also be attributed to nucleotide sequence variation [1]. Translation rate varies notably with the coding sequence [2, 3] due to differences in the codon-specific rates of initiation and elongation. It is already well known that single variations in codons for the same amino acid can strongly influence the overall expression process. In particular, these variations may be of the utmost importance to heterologous gene expression. The impact of single variations has been demonstrated for the structural folding of mRNA [4], with possible influences on mRNA degradation and/or initiation of translation. Even protein secondary structures are in some cases correlated with specific codon usage [5]. This effect may be caused by the impact of different translation accuracies for specific codons. Because of all of these impressive examples, codon optimization is an important issue for recombinant gene expression. The high number of dimensions of the parameter space justifies attempts to support this difficult design task by mathematical modeling and subsequent model-aided optimization of the gene sequence.

There are further interesting biotechnical applications which should benefit from such a sequence-oriented modeling. New challenges, for example, arise in the pursuit of vaccination with DNA and RNA. In particular, a sufficient expression level as well as the biological functionality and the tailored stability of the RNA are important issues which might be influenced by codon usage. Predictive models taking into account the variation of specific codons could support this difficult design task.

Since the final objective of the approach – the dynamic simulation of the parallel formation of the entire proteome under the in vivo conditions of a living cell – is still some way away, it is more realistic to envisage applications within the more simple area of in vitro protein biosynthesis. These systems allow us to study particular aspects of transcription and translation, such as the dynamic behavior in response to system perturbations. The main advantages of this approach come from the reduced complexity of these systems in comparison to a growing organism and their convenient accessibility. Additionally, however, the cell-free protein biosynthesis process has many interesting and promising applications which require a more systematic investigation of the bottlenecks in the productivity and stability of the system. Apart from model validation, the integrated model is therefore used to study the interrelatedness of the system components involved and to remove any bottlenecks in the underlying cell-free protein synthesis process. The challenge is again to improve the performance of the system with the aid of model-based optimization strategies.

Our development of the rigorous dynamic model for sequence-oriented gene expression is an attempt to aggregate existing biological knowledge of the individual reaction steps. The advantage of such an approach is that many of the kinetic parameters for the individual reactions can be taken from the literature. Accordingly, the review paper addresses the following issues: (1) transcription, (2) RNA degradation (3), translation and model validation with the aid of experimental observations from cell-free biosynthesis. These topics will, however, be preceded by a comprehensive overview of various strategies used in the dynamic modeling of gene expression.

2
Modeling Methodologies Utilized in the Simulation of Dynamic Gene Expression

In order to provide a basis for model selection, in the section we review the most important modeling strategies related to the dynamics of gene expression. We also briefly address the trade-offs associated with the different approaches. As with gene network modeling, there are two basic approaches used to model the dynamics of single gene expression – the “logical” or “Boolean” method, and the “dynamic-systems” method that uses ordinary differential equations. More detailed reviews of the literature will be presented in the context of the individual modules of transcription, mRNA degradation and translation.

2.1 Discrete Dynamic Systems

Discrete models are rule-based, where a stochastic event either takes place or does not according to the probability for this event to occur. Simple rules define a flow or change of state. Their computational efficiency makes these models particularly attractive when applied to large systems. On the other hand, a major drawback arises from the fact that only finite changes from one discrete state to another can be monitored using such models.

Discrete models were used extensively to describe protein biosynthesis mathematically. Gordon [6] modeled the states of ribosomes bound to a single mRNA in vector notation and computed polysomal size-distributions for various parameter sets. In this model, conditional probabilities for each discrete event, such as translation initiation, elongation, and termination, were chosen arbitrarily using Monte-Carlo simulations. Vassart et al. [7] extended the earlier approach to cover ribosome dynamics for a fixed number of mRNA molecules by using a matrix representation (Fig. 1). In this figure, rows denote mRNA molecules, columns indicate mRNA segments. The number given in each matrix element indicates the position (relative to each segment) that is covered by a ribosome. The model was later refined [8, 9] and used to investigate various aspects of ribosomal translation. Harley et al. [10] simulated protein synthesis under severe amino acid limitations. Menninger [11] considered the impact of an erroneous tRNA selection. Liljenström and von Heijne [12] accounted for variable elongation rates, and Bagnoli and Liò [13] differentiated between codons and tRNA diversity.

A similar discrete model to the one by Vassart et al. [7] was developed by Li et al. [14]. However, these authors achieved a deterministic model by

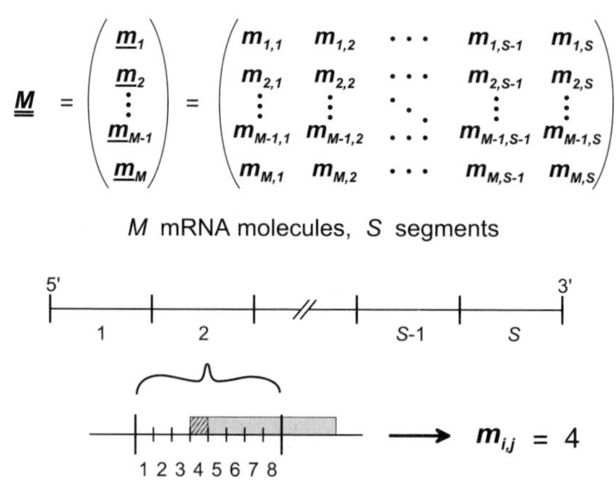

Fig. 1 Discrete modeling of ribosome states. Matrix element $m_{i,j}$ denotes the position of a ribosome (gray-shaded rectangle) bound to segment j of mRNA i

assigning fixed time intervals to the different states a system variable can take. Singh [15] developed a stochastic model to simulate the size distribution of polyribosomes and mRNA degradation. Much later, the same author combined his earlier model with a Markov model [16], which provides the necessary probabilities for state transitions. Carrier and Keasling [17] applied a stochastic model for studying mRNA degradation mechanism embedded in prokaryotic gene expression.

Another discrete modeling approach was taken by Gouy and Grantham [18]. These authors derived a probabilistic model of the tRNA cycle that simulates the behavior of single molecules. Such an approach makes it necessary to consider the spatial three-dimensional distribution of state variables. Although computationally expensive, these models are valuable, in particular, for systems that contain state variables in very small numbers.

2.2 Continuous Modeling

Continuous models take the form of (nonlinear) differential and algebraic equations and thereby allow us to trace the continuous changes in system variables, including their intermediate states. These models have been formulated by treating the rates of transcription, translation and mRNA degradation in a black-box approach. In these models, state variables (like concentrations of genes and mRNA) enter the kinetic expression in a linear fashion. First-order reaction rates are thus obtained with respect to these state variables (see Fig. 2). Black-box models are widely used where there is only a limited amount of knowledge available about a particular reaction. When the main emphasis of an investigation is placed primarily on the model structure (the connecting links between the state variables), it may be worthwhile accepting a reduced level of detail in the description of the reaction kinetics. In this context, black-box models have been considered for structured gene expression systems [19–21], and also for stability analysis [22, 23]. Black-box models are also attractive for large reaction networks, such as in the study of pharmacokinetics in gene therapy (Ledley and Ledle [24]).

Probably the most compelling advantage of unstructured models is their simplicity. Frequently, an analytical solution exists for these models, making numerical integrations obsolete. Only a single parameter is needed for each first-order reaction to fully describe the kinetics. However, this benefit also contributes the most severe limitation of unstructured models, that further rate-determining factors are neglected. For gene expression models based on the black-box assumption, this means that they miss out on the impact of cellular regulation, denoted by the variety of synthesis rates and degradation rates observed. Model parameters thus need to be estimated experimentally and separately for each protein product, which imposes large constraints on the predictive capacities of such models.

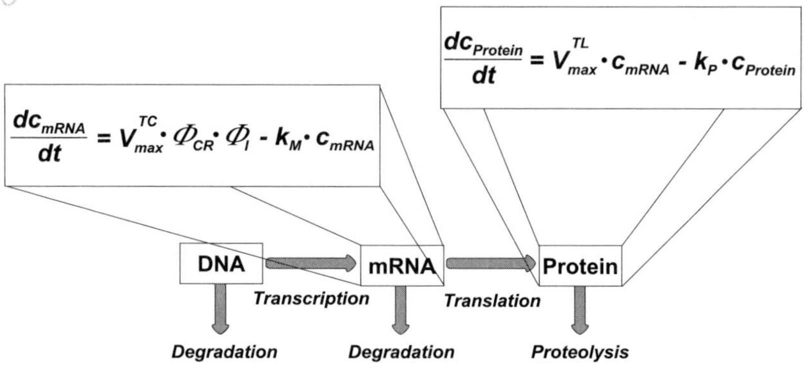

Fig. 2 Example of the use of unstructured modeling for representing gene expression. Material balance equations are provided for concentrations of both mRNA and protein. Symbol V_{max} denotes the maximum rate of both transcription (TC) and translation (TL), respectively. Φ_I is the defined as the fraction of free operator to total operator genes, while Φ_{CR} denotes the fraction of occupied promoters to the total number of promoter genes. Thus, these efficiency factors may themselves represent functional dependencies on the concentrations of both the repressor and operator regions. Constants k_M and k_P are first-order degradation constants

With more knowledge becoming available about reaction mechanisms, unstructured gene expression kinetics may be refined appropriately in order to tackle this problem. The initial idea goes back to a formalism provided in the 1970s by Aiba and co-workers [25], who derived an efficiency factor for both transcription and translation. These factors express a functional dependency on the concentration of regulatory components and may be multiplied by the respective maximum rate to modulate the conversion rate (see Fig. 2). Model expansions leading to genetically structured models were given by Bailey and co-workers (Lee and Bailey [26]; Chen et al. [27]).

More sophisticated continuous models have been developed for simulating DNA replication [28–30]. Gerst and Levine [31] developed a deterministic model that uses differential equations to describe the dynamics of polyribosomes. However, these authors omitted the impact of sterical interactions among translating ribosomes. In a steady-state analysis, Godefroy-Colburn and Thach [32] investigated the effect of mRNA competition on regulating translation rates. These authors further considered the case where translation initiation is blocked by ribosomes that are already bound within the initiation site.

A continuous model for reversible polymerization processes on a template was developed by the working group of Gibbs [33–35]. Characteristic to their approach is the step-wise travel of a catalyst along the template, whereby a monomer is linked to a nascent product chain at each step. The biopolymer synthesis considered an analogy to the physical problem of cooperative diffusion along a one-dimensional lattice [33]. Mass transfer rates for successive monomer addition were derived on the basis of the fractional loading of each template site (MacDonald et al. [34]). The same model structure was later extended to describe the impact of mRNA secondary structure on the overall translation rate (von Heijne et al. [36, 37]). Under simplifying assumptions regarding the original model, it was moreover possible to reduce the number of differential equations to a single one (Heinrich and Rapaport [38]). This model reduction holds only for the special situation if translating ribosomes are uniformly distributed over the length of a mRNA (including the termination site), and when they all propagate at the same specific rate.

Heinrich and Rapaport [38] performed a transition from fractions to molarities and included a balance for total ribosomes. These authors were the first to provide time-dependent solutions to a translation model. They also treated a system of two competing mRNAs, which differed in their rate constants for translation initiation.

Apart from the above continuous models, gene expression has been modeled as an autocatalytic relaxation process (Chela-Flores et al. [39]). Mahaffy [40] lumped all steps involved in both transcription and translation together to form a time delay until the full-length protein is assembled. In order to study the effects of clustering of low-usage codons (rare codons) as a function of their position along the mRNA and their impact on protein production rate, Zhang et al. [41] developed a prokaryotic translation model consisting of algebraic equations. Their model illustrates the positions of ribosomes on a mRNA and their residence times at different codons. The model is also capable of including interactions among polyribosomes. Götz and Reuss [42] modeled time delays in microbial growth by considering the polymerization reaction of ribosome synthesis. In a recent study by Drew [43], prokaryotic protein synthesis was modeled on the basis that transcription initiation rate is modulated by various states that the polymerase binding site can take (such as being activated or repressed). Probabilities for the different states of DNA were represented by a Markov model, and their time evolutions were given by a continuous black-box model. However, no polyribosomes and hence no queueing effects were considered.

3
Transcription

The sequence-oriented modeling of transcription has been elaborated in detail by Arnold et al. [44]. Given the need to integrate the corresponding module into a holistic model of gene expression, the structure of this module will be subsequently revisited in a condensed form.

The reaction scheme displayed in Fig. 3 was derived according to the common understanding of the transcription mechanism. T7 RNA polymerase (T7

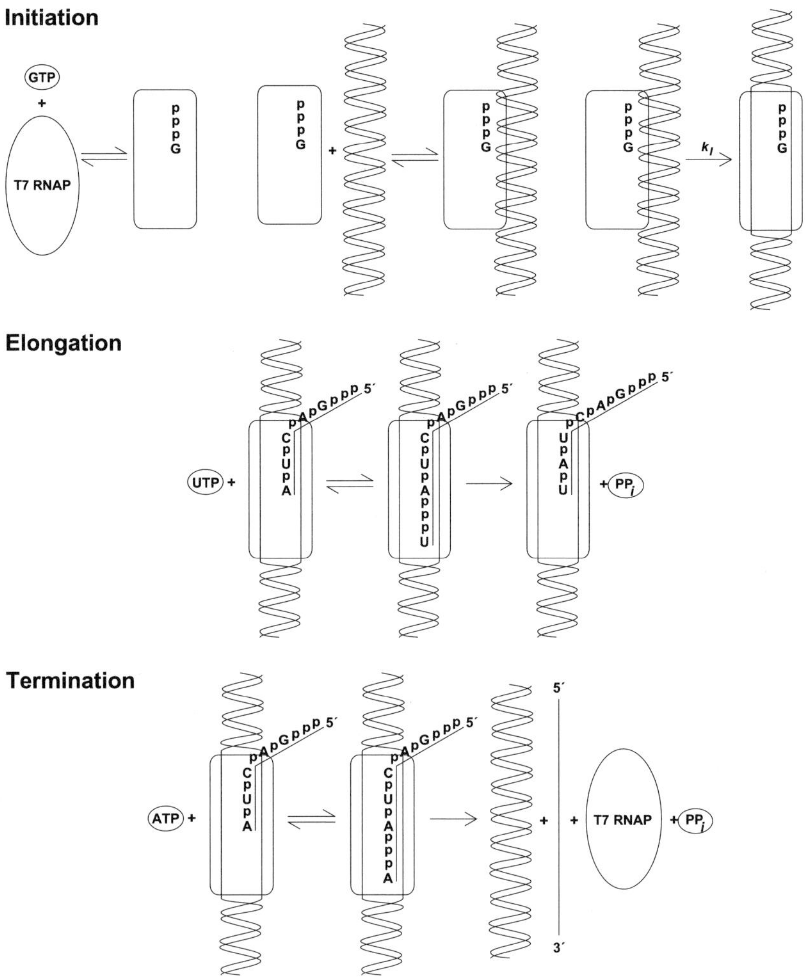

Fig. 3 Principle scheme for transcription by T7 RNA polymerase

RNAP) was chosen as a model system and also employed for the experimental validation of the model (Arnold et al. [44]).

Initiation. GTP is the initiator nucleotide. A random order of binding of T7 RNAP to the promoter, *D*, and GTP is possible. T7 RNAP is highly specific to its promoter, with a binding constant for promoter association of 1.0×10^8 M^{-1} versus a binding constant of nonpromoter association of 2.1×10^4 M^{-1} [45]. Nonspecific binding to DNA is neglected.

Elongation. Nucleotide association to the transcription complex of T7 RNAP, DNA, and RNA$_j$ is independent of neighboring nucleotides of the DNA sequence. The rate constant, k_{TCE}, denotes an irreversible translocation step, during which one molecule of inorganic pyrophosphate is released.

Competitive inhibition. Nucleotides and inorganic pyrophosphate competing with the binding of cognate substrate nucleotide are allowed to bind to freely dissolved T7 RNAP, to the enzyme-promoter complex, and to the elongating enzyme. The error frequency for transcription is negligible, with a reported probability of 10^{-5} [46].

Termination. The processes involved in transcription termination are combined into one irreversible reaction step, during which the fully synthesized RNA product is released.

The kinetic model developed inherently assumes that the system has settled into a pseudo-steady state. While the validity of this assumption has not been deliberately tested in this study, there is some support to be found in the literature. Guajardo et al. [47] observed a simultaneous linear increase in the concentrations of different RNA species (run-off, fall-off, and abortive transcripts). This increase continued at levels proportionately above nonlimiting substrate levels. These results provide strong evidence that steady-state synthesis was indeed achieved within the short time frame of a few seconds. Thus, the period of pre-steady state kinetics appears to be negligible when this model is applied to simulate several minutes of process time.

3.1 Reaction Kinetics

Using Fig. 3, the rate of total RNA synthesis, V_{TC}, by T7 RNAP under in vitro conditions has been derived mathematically to give the following functional dependence on the concentrations of NTP, total promoter (C_{D}), and inhibitory byproduct PPi:

$$V_{\mathrm{TC}} = \frac{V_{\mathrm{TC}}^{\max}}{D} \tag{1}$$

with

$$D = 1 + \sum_{j=1}^{N} \frac{K_{\mathrm{M,NTP},j}}{C_{\mathrm{NTP},j}} \left(1 + \frac{C_{\mathrm{PPi}}}{K_{\mathrm{I,PPi}}} + \sum_{i=1, i \neq j}^{N} \frac{C_{\mathrm{NTP},i}}{K_{\mathrm{I,NTP},i}} \right) + \frac{K_{\mathrm{M,D}}}{C_{\mathrm{D}}} \left[1 + \frac{K_{\mathrm{G}}^{\mathrm{I}}}{C_{\mathrm{GTP}}} \left(1 + \frac{C_{\mathrm{PPi}}}{K_{\mathrm{I,PPi}}} + \sum_{j=1}^{N-1} \frac{C_{\mathrm{NTP},i}}{K_{\mathrm{I,NTP},i}} \right) \right] .$$

Model parameters used in this rate equation are themselves composed of rate constants for elementary reaction steps and association constants for substrate binding. Their mathematical expressions are shown in Table 1. Importantly, the derived transcription kinetics include genomic sequence information in terms of transcript length, transcript composition, and the rate constants for initiation, elongation, and termination of RNA polymerization. These rate constants are vector-specific and vary with the consensus sequence of regulatory elements like the sites of promoter binding and transcription termination.

Neglecting substrate competition, the denominator of Eq. 1 simplifies to

$$D = 1 + \sum_{j=1}^{N} \frac{K_{\mathrm{M,NTP},j}}{C_{\mathrm{NTP},j}} \left(1 + \frac{C_{\mathrm{PPi}}}{K_{\mathrm{I,PPi}}}\right) + \frac{K_{\mathrm{M,D}}}{C_{\mathrm{D}}} \left[1 + \frac{K_{\mathrm{G}}^{\mathrm{I}}}{C_{\mathrm{GTP}}} \left(1 + \frac{C_{\mathrm{PPi}}}{K_{\mathrm{I,PPi}}}\right)\right] . \quad (2)$$

Material balances for a batch-wise transcription employing T7 RNAP may be formulated for total RNA concentration, all substrate nucleotides individually, and for inorganic pyrophosphate, to achieve:

$$\frac{\mathrm{d}C_{\mathrm{RNA}}}{\mathrm{d}t} = \sum_{i=1}^{R} V_{\mathrm{TC},i} \quad (3)$$

$$\frac{\mathrm{d}C_{\mathrm{NTP},j}}{\mathrm{d}t} = -\sum_{i=1}^{R} f_{j,i} n_i V_{\mathrm{TC},i} \quad \text{for} \quad j = 1 \text{ to } N \quad (4)$$

$$\frac{\mathrm{d}C_{\mathrm{PPi}}}{\mathrm{d}t} = \sum_{i=1}^{R} \frac{n_i - 1}{n_i} V_{\mathrm{TC},i} \, . \quad (5)$$

Table 1 Estimated kinetic parameters for in vitro transcription by T7 RNA polymerase using plasmid pT3/T7luc

Parameter		Unit	Value
$V_{\mathrm{TC}}^{\max}$	$k_{\mathrm{TC}} C_{E,t}$	μM/min	188
$K_{\mathrm{M,D}}$	$\frac{k_{\mathrm{TC}}}{k_{\mathrm{TCI}} K_{\mathrm{D}}}$	nM	6.3
$K_{\mathrm{M,ATP}}$	$n_{\mathrm{A}} \frac{k_{\mathrm{TC}}}{k_{\mathrm{TCE}}} K_{\mathrm{A}}$	μM	76
$K_{\mathrm{M,CTP}}$	$n_{\mathrm{C}} \frac{k_{\mathrm{TC}}}{k_{\mathrm{TCE}}} K_{\mathrm{C}}$	μM	34
$K_{\mathrm{M,GTP}}$	$k_{\mathrm{TC}} \left[\frac{n_{\mathrm{G}} - 1}{k_E} K_{\mathrm{G}} + \frac{1}{k_{\mathrm{TCI}}} K_{\mathrm{G}}^{\mathrm{I}}\right]$	μM	76
$K_{\mathrm{M,UTP}}$	$n_{\mathrm{U}} \frac{k_{\mathrm{TC}}}{k_{\mathrm{TCE}}} K_{\mathrm{U}}$	μM	33
$K_{\mathrm{I,PPi}}$		μM	200
$k_{\mathrm{d,TC}}$		min^{-1}	0.014

Parameter $f_{j,i}$ indicates the molar fraction of base j contained in transcript i. For more detailed information, particularly regarding the estimation of parameters from experiments, including their biochemical interpretation in terms of incorporation of sequence data, the reader is referred to the original paper.

3.2 Discussion of the Transcription Model

Although other kinetic models have been developed in the past to describe the dynamics of transcription, apparently none of these models has placed enough emphasis on a systematic mechanistic model derivation, which could have ultimately led to an expression for the transcription rate in terms of specific DNA characteristics. The particular novelty of this approach arises from the fact that the developed transcription model attempts to make use of genomic sequence data and annotated information in order to predict the transcript synthesis rate. Sequence data incorporated into the model include (a) the explicit locations of initiation and termination sites, and (b) the nucleotide sequence in-between these sites. From these two pieces of information, the lengths of RNA transcripts to be synthesized and their nucleotide compositions are readily calculated. When the specific recognition sequences of initiation and termination sites are also known and have been tabulated with their corresponding rate constants, then these parameters can be conveniently selected from such a library and used to simulate the transcription rate. A large collection of transcription factor recognition sites and annotated information concerning their binding properties is accessible in such databases, such as TRRD (Kolchanov et al. [48] and TRANSFAC (Wingender et al. [49]).

The general formulation of lumped model constants in terms of sequence-oriented parameters allows us to enter the respective information for each investigated system and thus greatly improves the range of applicability of this model. From among the model parameters, the maximum transcription rate, V_{TC}^{max} was selected to undergo a more detailed examination with respect to how it is influenced by the genomic sequence (Arnold et al. [44]).

The model developed may be used in the dynamic simulation of mRNA synthesis rate as part of (both in vivo and in vitro) recombinant protein production systems employing T7 RNA polymerase and the investigated transcription initiation and termination sites. In combination with a mathematical model of mRNA degradation, the transcription model could serve as a basis for system design.

The structural similarities identified between nucleic acid polymerases [50] may also provide an indication of the mechanistic similarities between these enzymes. It would thus be interesting to test the transferability of this model in order to describe mRNA synthesis rate by a RNA polymerase other than from bacteriophage T7. In such an approach, obviously the respective kinetic parameters specific to this particular RNA polymerase need to be known.

Additional kinetic features, such as the involvement of transcription factors for example, are at present not included in this model. With the current model formulation, however, it should in principle be possible to add further mechanistic properties. In this context, knowledge about binding constants for transcription factor binding is necessary. Modeling would then greatly benefit from studies providing these binding constants, either obtained from experimental detection, or alternatively from theoretical derivation on the basis of thermodynamic constraints (Kolchanov et al. [48]).

4 Prokaryotic mRNA Degradation

4.1 Introduction

Messenger RNA (mRNA) plays a central role in gene expression regulation, since this molecule constitutes the connecting link between genetic information and ribosomal protein synthesis. In general, protein expression rates are correlated with transcript levels and the efficiency with which these transcripts are translated. The effective mRNA concentration results from a superposition of transcript synthesis and degradation through ribonucleolysis.

Functional half-lives of mRNA typically range from 1 to 5 min in prokaryotes [51, 52], reach up to 25 min in yeast, and up to 16 hours in mammalian cell cultures [1, 53, 54]. While a fast mRNA turnover is a vital requirement for the cell to be able to quickly adapt to environmental changes, a sufficient mRNA stability is also necessary for the successful application of recombinant DNA technologies.

The mechanism for mRNA degradation in *E. coli* is commonly believed to proceed from $5'$ to $3'$ of the mRNA and involves the so-called degradosome. This aggregate of multiple enzymes contains both endonucleases and exonucleases, and is moreover capable of unwinding mRNA secondary structures [55–57]. RNase E, a main component of the degradosome, selectively recognizes endonucleolytic cleavage sites that are characterized by an enrichment of adenine (A) and uracil (U). The study by McDowall et al. [58] suggested that these sites are determined by their A/U-content rather than by the particular order of the nucleotide. RNase E was shown to associate to the $5'$-end of the mRNA when initiating the degradation process [59]. RNA secondary structural elements like stem-loops at the $5'$-terminus constitute sterical obstacles to the association of the degradosome. Stem-loop structures may also affect degradosomal migration along the mRNA in the search for endonucleolytic cleavage sites and may further impair the catalytic step of endonucleolytic cleavage itself.

The exonuclease polynucleotide phosphorylase (PNPase) contained also in the degradosome degrades the RNA fragments resultant from endonucleolytic cleavage. According to common belief, PNPase operates in the 5′-direction and remains attached to the mRNA molecule until the latter is fully digested [60].

The importance of the degradosome as a key player in bacterial mRNA degradation has been further emphasized as new enzymes have been found to participate in degradosome catalysis. After the initial degradosome binding to the mRNA at its 5′-terminus [60], an alternating sequence of degradosome propagation, scanning the mRNA for endonucleolytic cleavage sites, and endonucleolytic cleavage followed by exonucleolytic digestion leads to the successive degradation of the mRNA molecule. The movement of the degradosome has been perceived as sliding along the mRNA following translating ribosomes [61]. Alternatively, degradosomes bound to 5′-tails of mRNA were considered to stochastically loop inwards and thus scan the mRNA for putative endonucleolytic cleavage sites (Carrier and Keasling [17]).

mRNA degradation rate is in many ways modulated by ribosomal translation. Binding of the 30S ribosomal subunit to the Shine-Dalgarno sequence in the vicinity of the 5′-terminal mRNA is capable of stabilizing *lacZ* mRNA [62]. Ribosomes bound to a mRNA may physically block degradosomes from entering the sites of nucleolytic cleavage [52]. Further, amino acid starvation was found to delay the degradation of *trp* mRNA [63, 64]. All of these examples share a modulation of ribosome densities along the mRNA in common. Thus, the spacing of translating ribosomes can be taken as an indicator of the level of mRNA protection [1, 65].

The rate of mRNA degradation is often modeled in terms of first-order kinetics, which are characterized by a single parameter, according to

$$\frac{dC_{\text{mRNA}}}{dt} = -k_{\text{d,mRNA}} C_{\text{mRNA}} \,. \tag{6}$$

Other mathematical models of mRNA degradation have been developed that treat the decay as a multi-step process. The stochastic model by Singh [15] envisions a random inactivation of the 5′-terminal mRNA by exonuclease activity, which is followed by a sequential mRNA degradation towards the 3′-end of mRNA. In a similar modeling approach, Rigney [66] considered a modulation of the degradation rate via the reaction of ribosome binding to the messenger. Further work in modeling mRNA degradation has been to mathematically describe the size distribution of a decaying mRNA population [67]. Moreover, in an attempt to discern between individual contributions to the overall observed chemical decay rate, Liang et al. [68] developed a deterministic model with two model parameters, one of which related to endonucleolytic cleavage and the other to exonucleolytic digestion.

Carrier and Keasling [17] provided a remarkably detailed mechanistic description of prokaryotic mRNA degradation. Their modeling approach took into account degradosome binding and ribosome protection, which were embedded within the context of both mRNA and protein synthesis. The modeling frame is based on the stochastic model by Vassart et al. [7], where, characteristically, the rates of the polymerization steps (initiation, elongation, and termination of both transcription and translation, respectively) are taken to be model constants.

While the model by Carrier and Keasling [17] was very valuable for discriminating against degradation mechanisms, such a non-deterministic model is limited in its capacity to predict mRNA decay rates. For improved general applicability, ideally covering universal mRNA products, a functional dependence of mRNA degradation rate on the specific transcript properties is essential.

In this study, we describe the first modeling approach to representing mRNA degradation kinetics that includes nucleotide sequence information. The model aims in particular to account for both endonucleolytic and exonucleolytic reaction steps encountered during the decay process, as well as to describe the interactions of mRNA degradation and ribosomal translation mechanistically.

4.2 Mathematical Model

4.2.1 Nomenclature

According to Fig. 4, mRNA base triplets are consecutively numbered in the 5′ to 3′-direction from $j = 1$ to J. The coding region stretches from the translational start site ($j = j_{R0}$) to codon $j = K$, just prior to the translational stop codon. It is assumed that $K \leq J$.

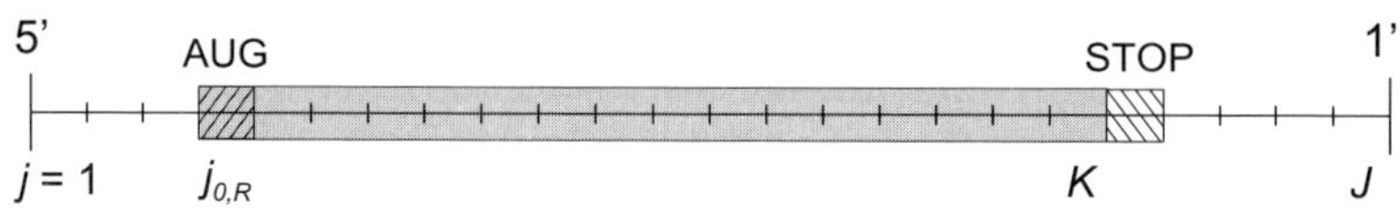

Fig. 4 mRNA with coding region (*gray-shaded*). The codons are numbered in the 5′ to 3′ direction from 1 to J by index j. $j_{0,R}$ designates the position of the translational start site, K the last codon of a coding region

Bound to a mRNA, a degradosome covers L_D base triplets at a time. A ribosome extends over L_R codons simultaneously. The catalytic center of bound degradosomes is located at m_D (with $1 \leq m_D \leq L_D$). The active center for protein synthesis is situated at position m_R of the ribosome (with $1 \leq m_R \leq L_R$). Both catalysts are believed to propagate into the same direction and one site at a time (see Fig. 5).

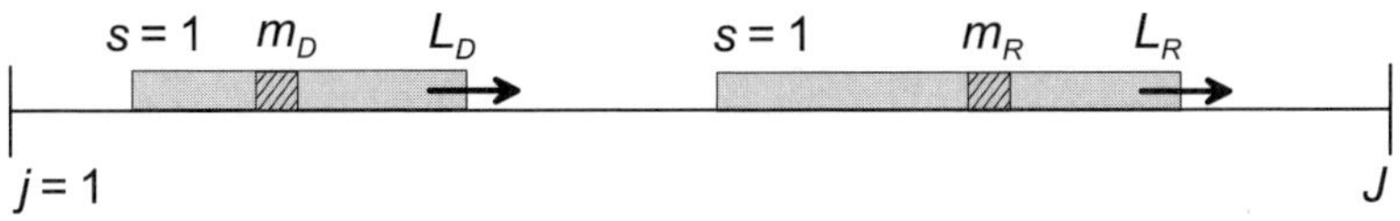

Fig. 5 Definition of states for two different types of catalysts bound to a template. The catalytic center of the bound degradosomes is located at m_{D}, the active center for protein synthesis at position m_{R} of the ribosome. The codons sterically covered by a catalyst are numbered in the 5′ to 3′ direction by s, from 1 to L_{D} in the case of degradosomes, and from 1 to L_{R} in the case of ribosomes

It is assumed that Z endonucleolytic cleavage sites exist for an arbitrary mRNA molecule (see Fig. 6). Position $z_1 = 1$ denotes the 5′-terminal base triplet of this mRNA. Base triplets j with $j \in \{z_2, ..., z_{Z-1}\}$ are characterized by an A/U-richness among their neighboring bases. In order to ensure full mRNA degradation, an additional cleavage site was introduced arbitrarily at the 3′-terminal base triplet ($j = J$).

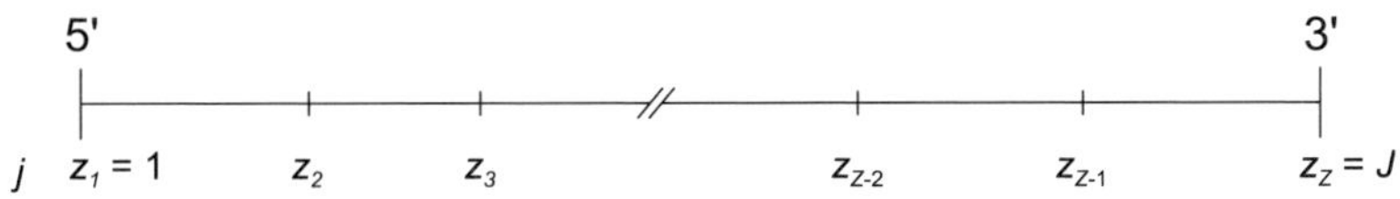

Fig. 6 mRNA with endonucleolytic cleavage sites. The codons are numbered in the 5′ to 3′ direction from 1 to J by index j. Cleavage sites are designated by z_i. Position $z_1 = 1$ denotes the 5′-terminal base triplet of this mRNA. Codons at position z_2 to z_{Z-1} are characterized by a A/U-richness among their neighboring bases. In order to ensure full mRNA degradation, an additional cleavage site was introduced arbitrarily at the 3′-terminal base triplet ($j = J$)

4.2.2
Reaction Scheme

The mechanism of mRNA degradation considered is conform with a typically observed 5′ to 3′-directed mRNA decay (Fig. 7). Ribosomes are assumed to be stripped off the mRNA before endonucleolytic cleavage takes place. The ordered series of reactions starts out with degradosome association to the 5′-end of substrate mRNA (step (1)). The degradosome travels along the mRNA until an A/U-rich stretch is recognized as an endonucleolytic cleavage site (step (2)). At this position, the degradosome will pause and endonucleolytically cut the mRNA. The newly-generated mRNA fragment is then transferred to the catalytic center of exonuclease activity (step (3)). Here, the fragment is successively degraded (step (4)). When this reaction is completed, the degradosome will continue its journey along the mRNA strand (step (5)) and will repeatedly undergo the stages of endonucleolytic and exonucleolytic digestion (steps (6) to (8)). The degradosome eventually arrives at the 3′-terminal end of the mRNA, and the remaining mRNA fragment is exonucleolytically degraded

(1)
(2)
(3)
(4)
(5)
(6)
(7)
(8)
(9)
(10)

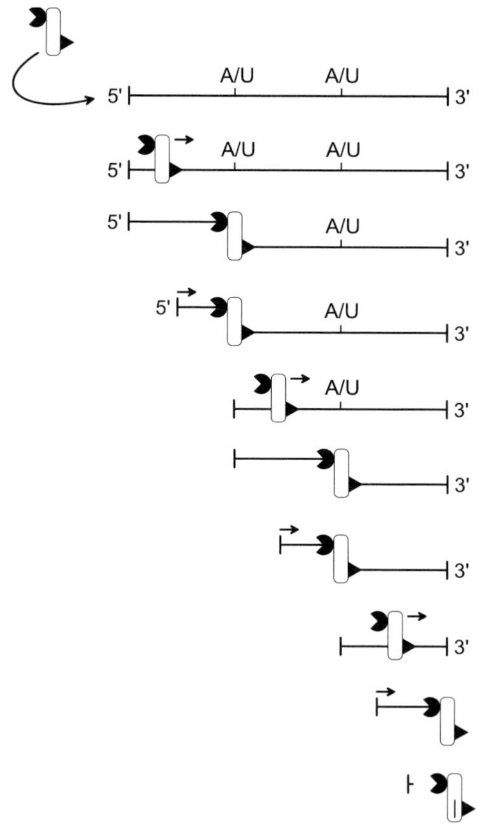

Fig. 7 Mechanism of 5′ to 3′-directional mRNA degradation

(step (9)). The decay process is terminated with the release of the degradosome (step (10)), which can subsequently reenter another degradation cycle.

4.2.3 Material Balancing

In the living cell (as well as under in vitro conditions), where mRNA molecules are constantly in the process of being generated while others are getting decomposed, it is difficult to envisage mRNA as a single type of species as opposed to a population of intermediates. From a modeling standpoint, such a high level of system complexity causes severe problems, in particular with increasing length of gene sequences. It appears impossible to track the fate of individual mRNA species by means of population balancing, unless further assumptions are made.

To arrive at a more practical formulation of system complexity, a site-specific state representation of state variables is chosen here. A reduction of system complexity is achieved through a projection of the entire mRNA population onto a single species of full-length mRNA. Material balance equations can now be derived for codon-specific variables, such as the total concentrations of each base triplet j, $\left(C_j^{\mathrm{M}}\right)$ with $1 \leq j \leq J$, and the concentrations of degradasomes $\left(C_{j''}^{\mathrm{D}}\right)$ and ribosomes $\left(C_j^{\mathrm{R}}\right)$ situated in j. These concentrations express averaged states with respect to the entire pool of each base triplet j.

For a system in which transcription initiation and translation initiation are switched off, the concentration of degradasome $\left(C_{j_{\mathrm{D0}}}^{\mathrm{D}}\right)$ bound to the association site at base triplet j_{D0} is affected by the rates of association and movement onto the next site, according to

$$\frac{\mathrm{d}C_{j_{\mathrm{D0}}}^{\mathrm{D}}}{\mathrm{d}t} = V_{\mathrm{D,ass}} - V_{\mathrm{D,mv},j_{\mathrm{D0}}} \,. \tag{7}$$

For all positions j with $j_{\mathrm{D0}} < j < J$ that do not coincide with an endonucleolytic cleavage site (i.e., $j \notin \{z_2, z_3, \ldots, z_{Z-1}\}$), the concentration of bound degradosomes is governed by the rate at which degradosomes enter this site and the rate of clearance:

$$\frac{\mathrm{d}C_j^{\mathrm{D}}}{\mathrm{d}t} = V_{\mathrm{D,mv},j-1} - V_{\mathrm{D,mv},j} \,. \tag{8}$$

Degradosome movement takes place until one of the endonucleolytic cleavage sites j is reached, with $j = z_i$ and $2 \leq i \leq Z$. At these particular sites, the degradosome will pause and adopt a state, here denoted by $C_j^{\mathrm{D}^*}$. In this state, an endonucleolytic cleavage reaction is considered to occur directly upstream of codon j, which generates a mRNA fragment of $(z_{i-1} - z_i)$ bases in length. The time-dependent change of concentration $C_j^{\mathrm{D}^*}$ with $j \in \{z_2, z_3, \ldots, z_{Z-1}, z_Z\}$ is given by

$$\frac{\mathrm{d}C_j^{\mathrm{D}^*}}{\mathrm{d}t} = V_{\mathrm{D,mv},j-1} - V_{\mathrm{D,endo},j} \,. \tag{9}$$

While the degradosome remains bound to the endonucleolytic cleavage site, the newly produced mRNA fragment is successively degraded by an exonuclease contained in the degradosome. The concentration of this degradosomal state is denoted by $C_j^{\mathrm{D}^*\mathrm{Frag}}$, with $j \in \{z_2, z_3, \ldots, z_{Z-1}, z_Z\}$, and changes with

$$\frac{\mathrm{d}C_j^{\mathrm{D}^*\mathrm{Frag}}}{\mathrm{d}t} = V_{\mathrm{D,endo},j} - V_{\mathrm{D,exo},j} \,. \tag{10}$$

After completion of the exonucleolytic digestion in position j with $j \in \{z_2, z_3, \ldots, z_{Z-1}, z_Z\}$, the degradosome will further propagate along the mRNA according to

$$\frac{\mathrm{d}C_j^{\mathrm{D}}}{\mathrm{d}t} = V_{\mathrm{D,exo},j} - V_{\mathrm{D,mv},j} \quad \text{for} \quad j \in \{z_2, z_3, \ldots, z_{Z-1}\} . \tag{11}$$

The material balance for degradosomes bound to the 3′-terminal base triplet is

$$\frac{\mathrm{d}C_J^{\mathrm{D}}}{\mathrm{d}t} = V_{\mathrm{D,exo},J} - V_{\mathrm{D,T}} \quad \text{for}\, j = J , \tag{12}$$

where symbol $V_{\mathrm{D,T}}$ used in Eq. 12 denotes the rate of degradation termination. Due to the fixed order of reaction steps that each degradosome needs to undergo in a degradation cycle, the pool of each base triplet j is governed only by the rates of endonucleolytic cleavage (given that transcription is stopped in this case). This means in particular that the concentration of base triplets can temporarily remain unaltered, even though it has been traversed by a degradosome. In this case, the $(z_{i-1} - z_i)$ base triplets in-between two consecutive cleavage sites, z_{i-1} and z_i change their states in parallel. In order to describe the time-dependent decrease of all J base triplets of a decaying transcript, it is thus sufficient to derive material balances for only Z selected base triplets (i.e., one for each mRNA fragment upstream of an endonucleolytic cleavage site, plus one balance for the 3′-terminal base triplet). The other concentrations of base triplets, C_j^{M} (with $1 \leq j < J - 1$ and $z_{i-1} \leq j < z_i$) can then be represented in terms of these reference states, i.e.,

$$C_j^{\mathrm{M}} = C_{z_{i-1}}^{\mathrm{M}} . \tag{13}$$

Due to Eq. 13, the time-dependent changes of all concentrations of mRNA base triplets can be described by the following Z material balances:

$$\frac{\mathrm{d}C_j^{\mathrm{M}}}{\mathrm{d}t} = - V_{\mathrm{D,endo},j} \quad \text{for} \quad j \in \{z_1, z_2, \ldots, z_{Z-1}\} \tag{14}$$

$$\frac{\mathrm{d}C_J^{\mathrm{M}}}{\mathrm{d}t} = - V_{\mathrm{D,T}} . \tag{15}$$

For a system comprising both mRNA degradation and ribosomal protein synthesis, additional balance equations need to be derived for the concentrations of mRNA-bound ribosomes. Under non-limiting growth conditions, metabolite pools (low molecular weight compounds) are approximately buffered, and the concentrations of cellular catalysts involved in ribosomal translation may be viewed to be constant. Therefore, these compounds are not balanced.

The material balance equations for the concentrations of ribosomes bound within the coding region of mRNA can thus be written as

$$\frac{dC^{R*}_{j_{R0}}}{dt} = V_{TLI,70SIC} - V_{TLI,IF2D} \quad \text{for} \quad j = j_{R0} \tag{16}$$

$$\frac{dC^{R}_{j_{R0}}}{dt} = V_{TLI,IF2D} - V_{TLE,j_{R0}} \quad \text{for} \quad j = j_{R0} \tag{17}$$

$$\frac{dC^{R}_{j}}{dt} = V_{TLE,j-1} - V_{TLE,j} \quad \text{for} \quad j_{R0} < j < K \tag{18}$$

$$\frac{dC^{R}_{K}}{dt} = V_{TLE,K-1} - V_{TLT} \quad \text{for} \quad j = K\,. \tag{19}$$

Symbol $C^{R*}_{j_{R0}}$ used in Eq. 16 refers to the concentration of 70S initiation complexes. After dissociation of initiation factor 2 (IF2), the concentration of ribosomes bound to the translational start site is given by $C^{R}_{j_{R0}}$. The concentration of ribosomes bound to position j is given by C^{R}_{j}.

4.2.4
Kinetic Rate Equations

Degradosome association was reflected by the rate expression

$$V_{D,ass} = k_{D,ass}\, q^{D0}_{j_{D0}}\, C^{M}_{j_{D0}}\,. \tag{20}$$

In Eq. 20, the total concentration of the base triplet (at which degradosome association takes place) is given by $C^{M}_{j_{D0}}$. The queueing factor, $q^{D0}_{j_{D0}}$, denotes the fraction of unoccupied 5′-binding sites. The derivation of this parameter is given in the Appendix (Sect. A.4). Queueing factors are by no means to be understood as model constants. Instead, they change dynamically, as the binding states of base triplets vary with time. According to their definition, queueing factors can take values between 0 and 1. Secondary structural features encountered in this region will render the rate constant, $k_{D,ass}$, for degradosome association. The value of this constant may also change with growth conditions because of variations in the free degradasome concentrations.

The stepwise one-directional diffusion of degradosomes along the mRNA is described by

$$V_{D,mv,j} = k_{D,mv}\, q^{D}_{j}\, C^{D}_{j}\,. \tag{21}$$

The rate of degradosome movement from base triplet j (with $j_{D0} \le j < J$) to position $j+1$ requires us to take into account sterical blocking by catalysts bound further downstream. Parameter q^{D}_{j} written in Eq. 21 denotes the probability of base triplet $j+1$ being unoccupied when a degradosome is located in j (see Appendix). The reaction rate for endonucleolytic cleavage comprises the steps involved in recognizing the site as a cleavage site, as well as the act of mRNA cleavage. The kinetics for this cleavage reaction at sites $j \in \{z_2, z_3, ..., z_{Z-1}, z_Z\}$ are represented by a first-order rate according to

$$V_{\mathrm{D,endo},j} = k_{\mathrm{D,endo},j} C_j^{\mathrm{D}^*} . \tag{22}$$

The rate constants, $k_{\mathrm{D,endo},j}$, may vary across all endonucleolytic cleavage sites. For convenience, this study treats all endonucleolytic cleavage sites the same, thus assigning the same parameter $k_{\mathrm{D,endo}}$ to any such sites. The total of all exonucleolytic steps can be summarized as

$$V_{\mathrm{D,exo},j,i} = \sum_{s=z_{i-1}}^{z_i} k_{\mathrm{D,exo},s}\, C_j^{\mathrm{D^*Frag}} , \tag{23}$$

with $j \in \{z_2, z_3, ..., z_{Z-1}, z_Z\}$ and $2 \leq i \leq Z$. The rate constant for exonuclease activity ($k_{\mathrm{D,exo},s}$) may differ with the type of base to be cleaved. It could also be influenced by sequence context. For example, each of the mRNA fragments may exhibit a unique secondary structural conformation. The unwinding of this structure, which is necessary during the process of an exonuclease reaction, would then lead to diverse rates of cleavage for each individual base in the exonuclease reaction. Although the model in its general form accounts for such differences, the rate constants for individual exonucleolytic cleavage steps will, in most cases, be unknown. For practical reasons, it is assumed further on that this parameter remains invariant with nucleotide sequence.

The termination rate of mRNA degradation, which occurs at the final base triplet ($j = J$) is assumed to obey a first-order rate law, according to

$$V_{\mathrm{D,T}} = k_{\mathrm{D,T}} C_J^{\mathrm{D}} . \tag{24}$$

In the case where mRNA degradation and ribosomal translation take place simultaneously, a two-step-mechanism for initiation of protein synthesis was considered. The first step is characterized by 70S initiation complex formation at the translational start site Eq. 25. In a second step, the dissociation of initiation factor 2 (IF2) is taken into account (Eq. 26).

$$V_{\mathrm{TLI,70SIC}} = k_{\mathrm{TLI,70SIC}}\, q_{j_{\mathrm{R0}}}^{\mathrm{R0}}\, C_{j_{\mathrm{R0}}}^{\mathrm{M}} \tag{25}$$

$$V_{\mathrm{TLI,IF2D}} = k_{\mathrm{TLI,IF2D}}\, C_{j_{\mathrm{R0}}}^{\mathrm{R}^*} \tag{26}$$

Symbol $C_{j_{\mathrm{R0}}}^{\mathrm{M}}$ stands for the concentration of base triplet j_{R0}. The kinetics for translation elongation and termination are given by Eqs. 27 and 28, respectively.

$$V_{\mathrm{TLE},j} = k_{\mathrm{TLE},j} q_j^{\mathrm{R}} C_j^{\mathrm{R}} \quad \text{for} \quad j_{\mathrm{R0}} \leq j < K \tag{27}$$

$$V_{\mathrm{TLT}} = k_{\mathrm{TLT}} C_K^{\mathrm{R}} . \tag{28}$$

The queueing factors $q_{j_{\mathrm{R0}}}^{\mathrm{R0}}$ and q_j^{R} used in Eqs. 25 and 26 denote the respective probabilities that base triplet j_{R0} and j are empty. These parameters are defined in the Appendix (Sect. A.4).

4.2.5 Model Reduction

When a less detailed description of states is acceptable, a significant reduction in the number of state variables can be achieved by merging groups of base triplets into one. Applying this method of model reduction, several consistency checks need to be performed. It is important to ensure that the reading frame of the coding sequence remains unaffected. Further, the influence of the new system representation on material balancing as well as the formulation of reaction kinetics and model parameters needs to be considered. In the case when translation elongation rates vary significantly in a codon-specific manner, material balancing of grouped base triplets and their states becomes more cumbersome (Sect. 5).

4.3 Parameter Identification for *lacZ* mRNA

The mathematical model of prokaryotic mRNA degradation presented in this study includes several model parameters that need to be identified in order for this model to become applicable for prediction purposes. These parameters are subsequently estimated for the example of *lacZ* mRNA. This well-studied gene has been chosen here for investigation because its mRNA is known to follow an exclusive 5′ to 3′ degradation pathway [68–70].

The sequence of the *lac*-operon was obtained for wild-type *Escherichia coli* K12 MG1655 from the European Molecular Biology Laboratory (EMBL, accession number AE000141). *lacZ* mRNA contains 3144 bases (= 1048 base triplets), considering the 5′ and 3′-ends reported earlier [71–73]. The coding region stretches from base triplets 14 ($= j_{\mathrm{R0}}$) to 1037 ($= K$), and is thus 1024 codons in length.

4.3.1 Half-lives of *lacZ* mRNA

Chemical half-lives of the 5′ and 3′-end of *lacZ* mRNA were reported for various growth conditions of *E. coli*. For a system in which translation initiation was inhibited, a half-life of 0.5 min was given for the 5′-terminal *lacZ* mRNA [74]. In the presence of an active translational machinery, the 5′-end is significantly stabilized and exhibits a chemical half-life of 1.9 min [68]. In the same study, the 3′-end of *lacZ* mRNA was also shown to be degraded with a half-life of 1.9 min, albeit after a one minute delay compared to the 5′-terminus. From these half-lives, the rate constants for exponential decay can be readily derived according to

$$k_{d,mRNA} = \frac{\ln 2}{t_{1/2}} . \tag{29}$$

4.3.2
Number of Endonucleolytic Cleavage Sites

Five primary endonucleolytic cleavage sites were verified experimentally for the 5′ and 3′-termini of *lacZ* mRNA [73, 75–77]. However, no such data exist for the major internal section of this mRNA. A close inspection of the identified cleavage sites reveals that these sites share in common a region of at least eight nucleotides in length and a content of both *G* and *C* of at the most 12.5%. Under the premise that this concept of identifying endonucleolytic cleavage sites also applies for the remainder of the *lacZ* mRNA, the nucleotide sequence has been scanned for putative endonucleolytic cleavage sites according to this search pattern. The outcome of this analysis is shown in Table 2. In addition to

Table 2 Estimated endonucleolytic cleavage sites for wild-type *lacZ* mRNA. Position indicates the start of an A/U-rich stretch relative to native full-length mRNA. Reported sites of cleavage are marked by a straight line. 1 = Subbarao and Kennell [76], 2 = Yarchuk et al. [77], 3 = Cannistraro et al. [71], 4 = McCormick et al. [73]

Position [nt]	G/C [nt]	Sequence [%]	Source
13	10.0	AU|AACAAUUU	1, 2
70	12.5	UUUU|AC|AA	1, 2
109	12.5	AACUU|AAU	1
419	10.0	|AUUUAAUGUU	1
461	7.7	AAUUAUUUUUGAU	
732	11.1	UUUAAUGAU	
814	11.1	UUUCUUUAU	
869	11.1	UGAAAUUAU	
1050	11.1	AUUGAAAAU	
1188	12.5	AACUUUAA	
1281	10.0	AAUAUUGAAA	
1531	0.0	AUAUUAUUU	
1599	10.0	AUCAAAAAAU	
1691	12.5	UAAAUACU	
1765	9.1	UGAUUAAAUAU	
2356	9.1	AUAAAAAACAA	
2586	10.0	UUAUUUAUCA	
2869	9.1	AAUUGAAUUAU	
3106	0.0	AAAAAU|AAUAAUAA	3, 4

the five experimentally-verified endonucleolytic cleavage sites for *lacZ* mRNA, 14 other such regions have been uncovered, which are proposed to function as RNase E recognition sites. Considering one additional cleavage site at the ultimate 3′-tail of *lacZ* mRNA, a total of 20 sites for endonucleolytic cleavage by RNase E were thus predicted. On average, one endonucleolytic cleavage site is suggested for about every 160 nucleotides.

4.3.3 Bounding Regions for the Parameter Range

The one minute time gap noted between 5′ and 3′-end degradation of *lacZ* mRNA in the presence of ribosomal translation denotes the cumulative time needed for each degradosome to travel along a full-length transcript molecule and to perform endonuclease and exonuclease activities during this propagation. This Δt imposes severe constraints on the mean duration of each of the reaction steps during mRNA degradation. The average time required for each step is given by the reciprocal of the corresponding rate constant. The sum of all time steps taken in the ordered process of mRNA degradation may thus be written as

$$\Delta t = \frac{J - j_{D0} - 1}{k_{D,mv}} + \frac{J - 1}{k_{D,exo}} + \frac{Z}{k_{D,endo}} . \tag{30}$$

Applying a limit case study, in which only one rate-limitation at a time is considered to occur, it is possible to estimate lower boundary values for each of the rate constants given above. That is, $k_{D,mv} \geq 17.5\ s^{-1}$, $k_{D,exo} \geq 17.5\ s^{-1}$, and $k_{D,endo} \geq Z/60\ s^{-1}$. The position for initial degradosome binding, j_{D0}, was taken to be equal to 1 in this rough estimation. The total number of endonucleolytic cleavage sites (Z) is not exactly known for *lacZ* mRNA. Using the method described in Sect. 2, $Z = 20$ sites in total were predicted for *lacZ* mRNA to be susceptible to RNase E attack. Hence, the rate constant for endonucleolytic cleavage ($k_{D,endo}$) is calculated to be greater than or equal to $0.3\ s^{-1}$.

4.4 Dynamic Simulation and Nonlinear Regression Analysis

4.4.1 Assumptions

1. Throughout the experiment, mRNA synthesis is completely prevented through blocking of transcription initiation.
2. The degradosome diameter approximates the physical dimensions of the ribosome: i.e., $L_D = L_R = 12$ codons [54, 78]. The reference states for degradosome and ribosome, respectively, are $m_D = m_R = 7$.

3. The 5′-end of *lacZ* mRNA hosts binding sites for both degradosome and ribosome association. As can be seen from Fig. 8, both sites overlap for the assumed ribosome and degradosome dimensions.
4. Parameter $k_{TLI,IF2D}$ was set to be equal to 0.8 s^{-1}, since this value was given for the effective frequency of translation initiation for wild-type *lacZ* mRNA under in vivo conditions [68].
5. In the case of *lacZ* mRNA, the average effective elongation rate of translating ribosomes, $(k_{TLE})_{eff}$, was reported to be 17.5 aa/s [68]. Sterical interactions among translating ribosomes are included in this value, i.e.,

$$(k_{TLE})_{eff} = q_j^R k_{TLE} \,. \tag{31}$$

6. Termination of mRNA degradation was assumed to be a non-limiting reaction step. The rate constant $k_{D,T}$ was arbitrarily selected to be equal to 50 s^{-1}.
7. Simulation starts out with full-length mRNA. No degradation products of mRNA are present at this time ($t = t_0$). The initial concentration of each base triplet, $C_j^M(t_0)$, with $1 \leq j \leq J$ was chosen to be 0.05 μM.
8. There are no degradosomes bound to full-length mRNA at the start of simulation. That is, $C_j^D(t_0) = 0$ μM for all j with $j_{D0} \leq j \leq J$.
9. For systems including ribosomal translation, the initial concentration of ribosomes bound to each codon j was taken to be equal to 2.3 nM.
10. Cell volume is regarded as being ideally mixed.

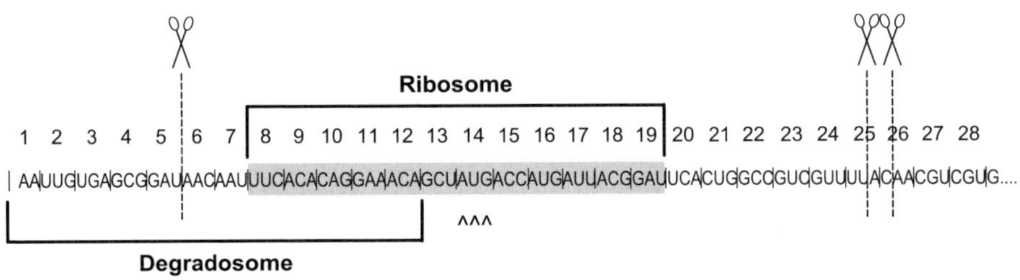

Fig. 8 For wild-type *lacZ* mRNA, the sites of degradosome and ribosome association overlap. Base triplets are sequentially numbered. The translational start codon is marked by arrows. Experimentally-verified endonucleolytic cleavage sites (see Table 2) are also indicated

4.4.2
Performance Index

With the measured chemical half-lives and the initial concentration of full-length mRNA, the time-dependent trajectory for 5′-terminal base triplets of mRNA (i.e., base triplet $j = 1$) can be written as

$$C_1^M(t) = C_1^M(t_0) \exp\left[-\frac{\ln 2}{t_{1/2}} \cdot t\right] \,. \tag{32}$$

The time-delayed first-order decay of the 3′-end of mRNA (i.e., base triplet $j = 1048$) is described by

$$C_{1048}^{\mathrm{M}}(t) = C_{1048}^{\mathrm{M}}(t_0) \tag{33}$$

for $t \leq \Delta t$, and for times greater than Δt by

$$C_{1048}^{\mathrm{M}}(t) = C_{1048}^{\mathrm{M}}(t_0) \exp\left[-\frac{\ln 2}{t_{1/2}} \cdot (t - \Delta t)\right] . \tag{34}$$

The goodness of fit was assessed by minimizing the sum of square relative errors. In these calculations, the setpoint concentrations of 5′ and 3′-terminal base triplets were taken at discrete time points from Eqs. 32 to 34, respectively, employing the reported chemical mRNA half-lives.

In addition to least squares fit analysis, the following parameters were monitored during simulation as model outputs in order to allow further assessment of system performance. The average spacing between ribosomes can be calculated from

$$d_{\mathrm{R}} = \frac{\sum\limits_{j=j_{\mathrm{R0}}}^{K} C_j^{\mathrm{M}}}{\sum\limits_{j=j_{\mathrm{R0}}}^{K} C_j^{\mathrm{R}}} . \tag{35}$$

The average spacing between degradosomes is given by

$$d_{\mathrm{D}} = \frac{\sum\limits_{j=j_{\mathrm{D0}}}^{J} C_j^{\mathrm{M}}}{\sum\limits_{j=j_{\mathrm{D0}}}^{J} C_j^{\mathrm{D}}} . \tag{36}$$

For times at which all concentrations of mRNA-bound degradosomes differ from 0, the average effective rate constant of degradosome movement can be obtained from

$$(k_{\mathrm{D}}, \mathrm{mv})_{\mathrm{avg}} = \frac{n_{\mathrm{c}}}{J - j_{\mathrm{D0}}} \sum_{j=j_{\mathrm{D0}}}^{J-1} \frac{V_{\mathrm{D,mv},j}}{C_j^{\mathrm{D}}} . \tag{37}$$

4.4.3 Parameter Estimation

In an attempt to identify model parameters with enhanced sensitivity, a sequential estimation procedure was applied. The identification of model parameters was initially carried out with a simplified state representation (see method described in Sect. 4.2.5). At first, the concentrations of mRNA and positional loadings were derived for every four adjacent base triplets ($n_c = 4$). The results of this analysis were compared at a later stage to results obtained using the model with full state representation (with $n_c = 1$).

4.4.3.1 Degradosome Association

From the degradation of 5′-terminal *lacZ* mRNA, when no translation was present, the rate constant of degradosome association, $k_{D,ass}$, was estimated to be 1.386 min^{-1}. The outcome from parameter estimation is given by the curve linking the black circles in Fig. 9. The parameter value identified for $k_{D,ass}$ was kept fixed throughout the subsequent estimation procedure.

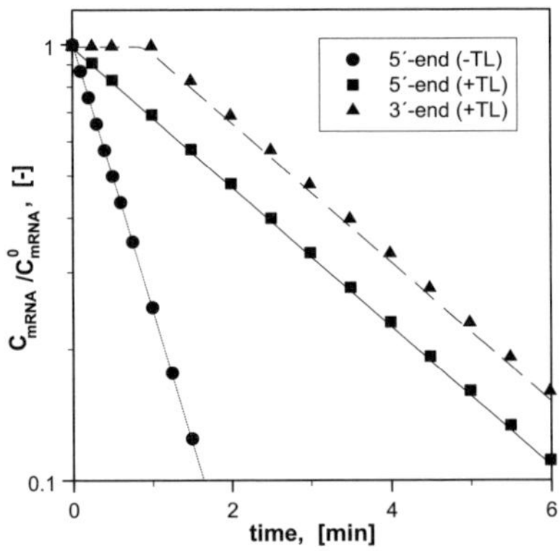

Fig. 9 Comparison of simulated versus experimental time course of terminal regions of *lacZ* mRNA. Relative concentrations are normalized with respect to their initial concentration. *Circles* denote the 5′-end of mRNA in the absence of translation. *Squares* and *triangles* refer to the 5′-end and the 3′-end of *lacZ* mRNA, respectively, in the presence of ribosomal translation. Experimental data were artificially generated from the mRNA half-lives provided by Schneider et al. [74] and Liang et al. [68]. Reduced model with $n_c = 4$

4.4.3.2 70S Initiation Complex Formation

Assuming that the increased mRNA stability due to translation is primarily caused by inhibited degradosome association, queueing factor q_{jD0}^{D0} can be estimated, as is outlined in the following. Using Eq. 20, the ratio of degradosome association rates of both systems with and without translation can be written as

$$\frac{(V_{D,ass})_{(+TL)}}{(V_{D,ass})_{(-TL)}} = \frac{\left(k_{D,ass}\, q_{jD0}^{D0}\, C_{jD0}^{M}\right)_{(+TL)}}{\left(k_{D,ass}\, q_{jD0}^{D0}\, C_{jD0}^{M}\right)_{(-TL)}} . \tag{38}$$

If the concentration of *lacZ* mRNA (C^{M}_{jD0}) and the rate constant for degradosome association ($k_{D,ass}$) are the same, whether translation prevails or is excluded, a difference in the rate of 5′-mRNA degradation between both systems would be reflected solely by q^{D0}_{jD0}. From Eq. 38, it is then possible to derive the following relationship:

$$\frac{(V_{D,ass})_{(+TL)}}{(V_{D,ass})_{(-TL)}} = \frac{\left(q^{D0}_{jD0}\right)_{(+TL)}}{\left(q^{D0}_{jD0}\right)_{(-TL)}} = \frac{(t_{1/2})_{(-TL)}}{(t_{1/2})_{(+TL)}} . \tag{39}$$

With Eq. 39, and assuming $\left(q^{D0}_{jD0}\right)_{(-TL)} \approx 1$ (in the case where no ribosomes are attached to mRNA), $\left(q^{D0}_{jD0}\right)_{(+TL)}$ is calculated to be 0.2632. This is a rough estimate under the assumption of unimpaired degradosome association. Parameter $\left(q^{D0}_{jD0}\right)_{(+TL)}$ was subsequently estimated from nonlinear regression analysis without the need for this simplification. The values taken by the queueing factor $\left(q^{D0}_{jD0}\right)_{(+TL)}$ are governed by the fractional occupancy of base triplets in the direct vicinity of the ribosome binding site. These fractional loadings are a primary result of the relative rates of translation initiation versus translation elongation. In the investigated example, parameters $(k_{TLE})_{eff}$ and $k_{TLI,IF2D}$ are fixed, as a result of experimental determination. The only model parameter left that can influence $\left(q^{D0}_{jD0}\right)_{(+TL)}$ is $k_{TLI,70SIC}$, which effectively determines the concentration of ribosomes attached to the ribosome binding site. Parameter $k_{TLI,70SIC}$ was estimated by fitting simulation results to the setpoint trajectory of 5′-terminal mRNA in the presence of translation (square symbols and solid line in Fig. 9). The rate constant of 70S initiation complex formation ($k_{TLI,70SIC}$) was thus determined to be 14.2 s^{-1}. Given this parameter value, the queueing factor for degradosome association $\left(q^{D0}_{jD0}\right)_{(+TL)}$ was found to be 0.2626, under pseudo-steady state conditions of mRNA degradation. The noted stability improvement of 5′-*lacZ* mRNA in the presence of translation could thus be explained exclusively by mRNA-bound ribosomes physically preventing access to the degradosome binding site.

4.4.3.3
Endonucleolytic and Exonucleolytic Cleavage, and Degradosome Movement

By fitting the simulated time course of the 3′-terminal base triplet of *lacZ* mRNA to its setpoint trajectory, the rate constant for endonucleolytic cleavage ($k_{D,endo}$) was estimated to be 2.6 s^{-1}. Estimates for the rate constants of exonucleolytic cleavage ($k_{D,exo}$) and degradosome movement ($k_{D,mv}$) were determined to be 680 nt s^{-1} and 95 nt s^{-1}, respectively. Figure 10 (triangles and dashed graph) illustrates the time dependency for 3′-*lacZ* mRNA obtained when using the identified parameter set in comparison to the experimentally-measured 3′-terminal base triplet concentration. A consistency check demonstrates that these estimated parameters are located well above their previously identified lower boundary values (see Sect. 4.3.3).

While the above parameter estimation was conducted with a simplified model exhibiting lower resolution of state variables ($n_C = 4$), the applicability of these parameters was subsequently tested by employing the model with full state representation ($n_C = 1$). When the same parameter set as estimated for the simplified model is applied to the full model, a mismatch between simulated time traces and experimental observation is noted for the system including ribosomal translation. The concentrations of mRNA base triplets are in this case proposed to be higher than in the experiment (see Fig. 10A). Nevertheless, the one minute time delay between 5′ and 3′-end degradation appears to be predicted correctly by the model. This finding, in combination with the similarity noted between both 5′ and 3′-terminal mRNA, suggests that it is mainly the degradosome association rate that is influenced by the effects of model reduction. When the rate constant for 70S initiation complex formation was then reevaluated, keeping $n_C = 1$, an improved fit between the simulated and the experimental time courses of both terminal mRNA base triplets was attained (see Fig. 10B). In this case, parameter $k_{TLI,70SIC}$ was estimated to be 4.3 s^{-1}. Thus, the degradosome association rate was indeed shown to be the most sensitive of the parameters of the mRNA degradation model to changes in state representation.

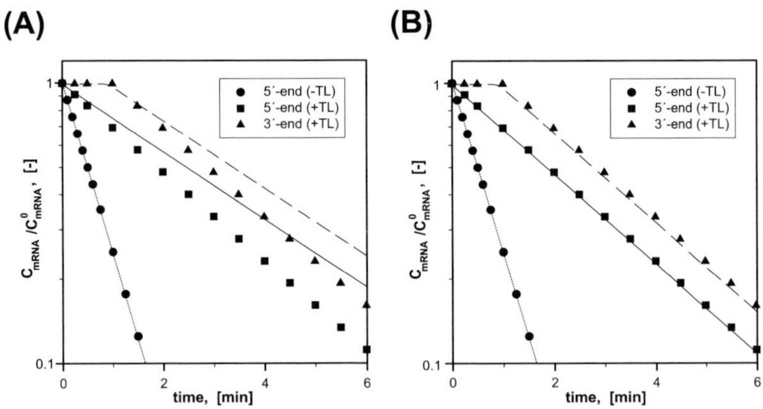

Fig. 10 Comparison of simulated versus experimental time course of both 5′ and 3′-ends of *lacZ* mRNA in the presence of ribosomal translation. Relative concentrations are normalized with respect to their initial concentrations. Experimental data were artificially generated from the mRNA half-life provided by Liang et al. [68]. (**a**) Full model with $n_C = 1$ and with model constants identified from the system with $n_C = 4$ (**b**) Full model with $n_C = 1$ and $k_{TLI,70SIC}$ equal to 4.3 s^{-1}

An explanation for the observed sensitivity becomes apparent from the implications of reduced state representation. For $n_c = 4$, ribosomes and degradosomes bound to mRNA cover a smaller number of positions at a time, namely 3 instead of 12 for the assumed case, while the physical dimensions of ribosomes, degradosomes and mRNA remain the same in either system representation. The queueing factor q_{jR0}^{R0} is then assembled for a smaller number of states of both ribosomes and degradosomes. These slight inaccuracies due to model simplification are shown to manifest themselves in an approximately threefold difference in the factor q_{jR0}^{R0}, the probability of the ribosome binding site being unoccupied. Under pseudo-steady state conditions, q_{jR0}^{R0} was 0.0345 for $n_c = 4$, while it was 0.1152 for $n_c = 1$. As a consequence of the above, parameter $k_{TLI,70SIC}$ was found to vary with the resolution of state representation.

Table 3 summarizes the effects of state resolution on characteristic quantities of the mRNA degradation model in combination with protein expression. In essence, it appears that merging base triplets leads to higher predicted concentrations of bound ribosomes, and consequently decreased values for queueing factors and average distances between ribosomes and degradosomes, respectively, and a reduced average effective rate of degradosome propagation.

Table 3 Model outputs from dynamic simulation and parameter identification. All quantities refer to quasi-steady state (qss) conditions of mRNA degradation in the presence of translation. Parameter n_c denotes the degree of codon refinement

Parameter	Unit	$n_c = 4$	$n_c = 1$
$\left(q_{jR0}^{R0}\right)_{qss}$	–	0.0345	0.1152
$\left(q_{jD0}^{D0}\right)_{qss}$	–	0.2626	0.2632
$\left(q_j^{D}\right)_{qss}$	–	0.8563	0.9747
$\left(k_{D,mv}\right)_{avg}$	codons/s	26.8	30.6
$k_{D,mv}$	codons/s	31.5	31.4
d_R	nt	110	150
d_D	nt	8600	9300
$\left(\frac{C_j^R}{C_j^M}\right)_{qss}$	–	0.11	0.02
$\left(\frac{C_{jR0}^{R*}+C_{jR0}^{R}}{C_{jR0}^{M}}\right)_{qss}$	–	0.73	0.65
$\frac{V_{D,ass}}{V_{TLI,70SIC}}$	–	0.01	0.01

The fractional occupancy of a particular codon j with respect to ribosome loading is given by the ratio of ribosome concentration bound to j and the concentration of this codon. That is, $C_j^{\mathrm{R}}/C_j^{\mathrm{M}}$. For $n_{\mathrm{c}} = 1$, this ratio is calculated to be 0.02 for all codons except for the initiation codon (see Table 3). In contrast, the translational start site (at $j = j_{\mathrm{R0}}$) is estimated to exhibit a higher ribosome loading (by a factor of 32.5, i.e., 0.65), supporting the notion that ribosomal binding to the translation initiation site functions as an effective mechanism to block upstream propagating degradosomes from entering the coding region. Finally, Table 4 lists the results from parameter estimation for the mRNA degradation model.

Table 4 Estimated parameters for the model of bacterial mRNA degradation employing *lacZ* mRNA in the presence of translation

Parameter	Unit	Value
$k_{\mathrm{D,ass}}$	s^{-1}	0.023
$k_{\mathrm{D,endo}}$	s^{-1}	2.6
$k_{\mathrm{D,exo}}$	nt s^{-1}	680
$k_{\mathrm{D,mv}}$	nt s^{-1}	95
$k_{\mathrm{TLI,70SIC}}$	s^{-1}	4.3

4.5 Discussion of the Submodel mRNA Degradation

The processes involved in mRNA degradation comprise an autonomous, separate modeling unit themselves. Nevertheless, care was taken to allow for the possibility of connecting the individual building blocks of a gene expression model in a modular fashion, in order to describe the performance of mRNA degradation embedded in prokaryotic gene expression. The level of detail with which the connected units (say, translation or mRNA synthesis) are represented may vary with the modeling task. For the purpose of parameter estimation, greater emphasis was placed in this study on modeling the mechanism of 5′ to 3′ mRNA degradation, while the kinetics of translation were treated in a simplistic manner. Apart from transcript length, the number and position of endonucleolytic cleavage sites, the steps involved in exonucleolytic digestion of mRNA, and the mechanism of mRNA protection through ribosomal translation were also included in the presented model.

As a direct consequence of the state projection, the model also describes situations where degradosomes are bound downstream of ribosomes, which is in contrast to the real system. Nevertheless, degradosomes and ribosomes bound to a particular codon j upstream of an endonucleolytic cleavage site do not get lost at the moment of cleavage. Instead they are – inherently in the model – redistributed within the remaining pool of base triplet j. Moreover, a reasonably sized set of state variables (maximally $3 \times J$) is obtained to characterize the concentrations of mRNA and bound ribosomes and degradosomes, respectively. The state vector is thus expected to be computationally more inexpensive than a system involving population balances. On the other hand, the projection procedure is clearly accompanied by a loss of information. In particular, conclusions about the loading pattern of individual mRNA molecules, their characteristic lengths, or the presence and integrity of their native $5'$ and $3'$-termini cannot be drawn using this model.

For the example of *lacZ* mRNA, it was possible to estimate the model constants of the presented mRNA degradation model. The general applicability of the identified parameter values to span a variety of mRNAs that follow a $5'$ to $3'$-degradation pathway, however, remains to be further exploited.

The mathematical model presented provides a framework for investigating the influence of ribosomal packing on mRNA protection against nucleolytic attack. An efficient translation initiation does not only lead to high protein expression rates. The results obtained in this study demonstrate that the efficiency of translation initiation also functions to control the stability of an mRNA transcript, when it conforms with the investigated degradation mechanism involving the degradosome. In this case,high fractional loadings of the ribosome binding site effectively function as a road-block to keep upstream degradosomes from accessing endonucleolytic cleavage sites that are contained within the coding region. Efficient translation initiation may thus lead to an autonomous amplification of protein expression rate.

The model takes into account the mechanism of mRNA protection by translating ribosomes both at the level of degradosome association (modulation of the accessibility of the degradosome binding site) and at the level of velocity of degradosome travel along the mRNA strand. Other than by sterical hindrance, inhibition by ribosomes that directly affects the rate of endonucleolytic cleavage is not accounted for by the model. Such a direct effect may arise from translating ribosomes that locally melt the secondary structural elements of mRNA during the process of peptide elongation. If not only sequence specificity, but also structural specificity is required to indicate an endonucleolytic cleavage site, such direct influence of ribosomes on the rate of endonucleolytic cleavage is conceivable. However, no evidence could be found in the relevant literature for any particular structure conservation role for the endonucleolytic cleavage sites recognized by RNase E.

Parameter estimation performed on the basis of lower system representation resolution can lead to an overestimation of queueing effects. A high sensitivity was observed for the association probabilities of both ribosomes and degradosomes dependent on the rate constant for 70S initiation complex formation.

Even if the concentration of bound ribosomes is in general expected to be orders of magnitude greater than the concentration of bound degrasosomes, it may become necessary – for technical reasons – to include the contribution of degradosomes in the queueing factor for ribosome elongation. In particular, with progressing mRNA degradation, the imbalance between the concentrations of bound ribosomes versus bound degradosomes will shift towards an increased fraction of bound degradosomes, which may then add significantly to the occupational status of a mRNA.

At a later stage of model development, the described reaction sequence for mRNA degradation may be further augmented by additional reactions. For example, it is conceivable that in future applications the particular effects of secondary structures that may be encountered both within the 5′ and the 3′-region of the mRNA, or that may form at intrinsic sites of mRNA when they are temporarily unoccupied by ribosomes, may be considered.

A highly detailed, sequence-oriented description of mRNA degradation has very important implications for practical application. It would be extremely valuable, if, with the aid of such models, pseudo-first-order rate constants for mRNA degradation could be inferred a priori for each different type of mRNA.

5 Prokaryotic Translation

5.1 Introduction

Ribosomal protein synthesis rates are known to vary with the protein product. It is generally accepted that codon composition, tRNA population and gene expressivity are strongly correlated [79]. The concentration of cognate tRNA is known to be positively correlated with the frequency of codon usage [80] Abundant proteins were found to be translated at a higher rate than rare proteins [81]. Elongation rate for two neighboring codons may be different by up to one order of magnitude [82]. Synonymous codons sharing the same cognate tRNA showed noticeably divergent elongation rates [83]. Variations in elongation rate have been attributed to differences in tRNA availability [84], and alternatively to the variability of binding constants for codon-anticodon interaction [83]. Codon context was considered to be insignificant when determining elongation rates [83]. An optimization of elongation rate along the mRNA can be accomplished through the preferential selection of synonymous codons matching those isoacceptor tRNAs that are abundant [82].

Queue formation among translating ribosomes has been demonstrated both in vitro [85], and in vivo, the latter in *Escherichia coli* during amino acid starvation [86]. Stalled ribosomes can cause a situation similar to that observed during a traffic jam in car traffic. A temporal hold-up of ribosomes, may result from downstream ribosomes scanning for the correct aminoacylated tRNA. Another example is the clustering of rare codons, which leads to more densly spaced ribosomes upstream and causes more distant spacing among ribosomes downstream of the cluster [41]. Such effects can lead to significantly lower rates of ribosomal movement than may be inferred from substrate availability, and could ultimately cumulate in a breakdown of protein synthesis, when at least one amino acid is missing.

Due to the central role of gene expression in cell metabolism, protein biosynthesis has been a major target of mathematical modeling. While individual features of translation have been modeled in great detail, a mechanistic model combining the majority of the key processes involved in one model is missing. This lack of a model is of particular importance in the pursuit of a thorough understanding of the molecular basis of ribosomal interactions.

In this study, a kinetic model of the prokaryotic translation process is developed that builds on the profound biomolecular knowledge gathered over the past decades. The model distinguishes between initiation, elongation, and termination of protein polymerization, and features the key catalysts enrolled in these reactions. Moreover, mutual interactions among ribosomes organized within a polysome structure are taken into account.

5.2 Initiation

In a complex multi-step process involving initiation factors IF1, IF2, and IF3, the binding of 30S ribosomal subunit to the initiator tRNA (fMet-tRNA^{M_f}), and their association to the ribosome binding site (RBS) of the mRNA are accomplished (see also Fig. 11).

5.2.1 Previous Modeling

Binding studies were carried out to determine the association constants for *E. coli* ribosomal subunit association and initiation factor binding at various ionic conditions [87–93]. Initial rate kinetics of translational initiation were derived from an in vitro system, by assuming a rapid equilibrium ordered mechanism for initiator tRNA binding to the 30S ribosomal subunit and the subsequent mRNA association [94]. Translation initiation kinetics were studied for *E. coli* derived systems using stopped-flow techniques to elucidate individual conformational changes and to measure the respective rates of elementary reactions [95, 96].

5.2.2 Reaction Scheme and Kinetics

The reaction scheme of bacterial translation initiation shown in Fig. 11 was derived from the above cited studies. The initiation process distinguishes the

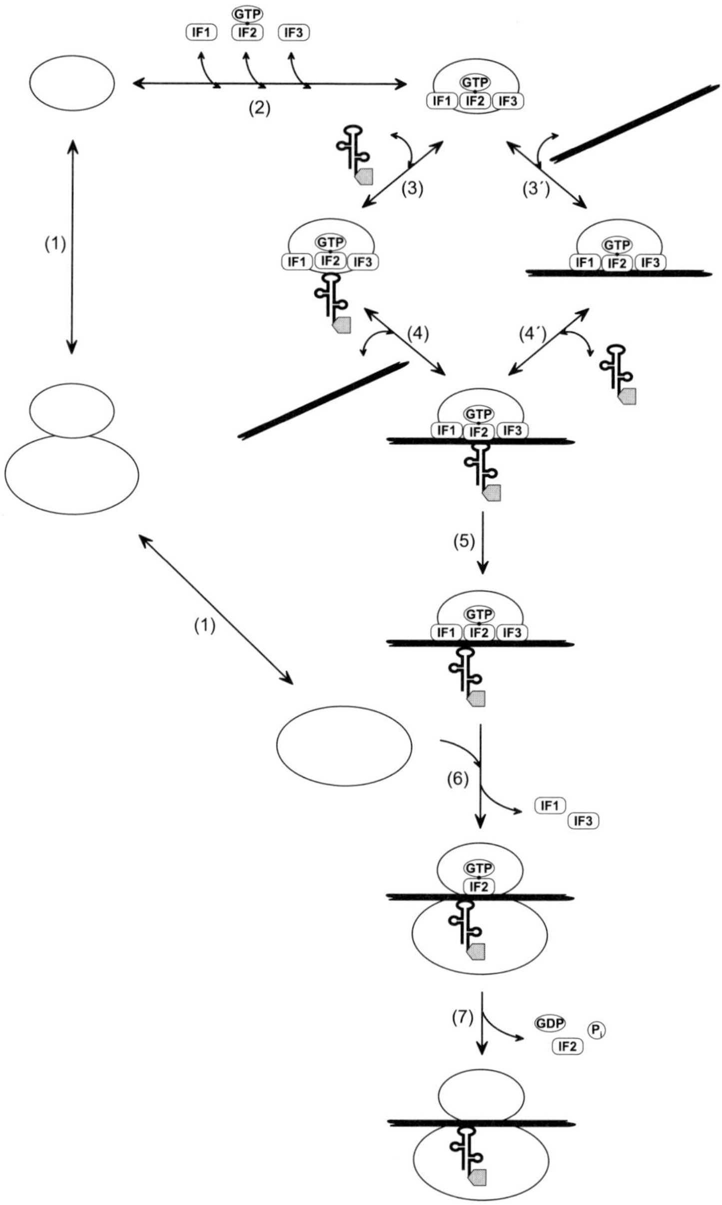

Fig. 11 Principle reaction scheme of prokaryotic translation initiation

steps of dissociation of ribosomal subunits (step (1)), association of initiation factors to 30S (step (2)), binding of ribosomal subunits to mRNA (steps (3) to (6)), and dissociation of IF2 from the mRNA-bound ribosome (step (7)).

Dissociation of Ribosomal Subunits

Under physiological conditions, the thermodynamic equilibrium of association of ribosomal subunits

$$30\text{S} + 50\text{S} \overset{K_{70\text{S}}}{\rightleftharpoons} 70\text{S} \tag{40}$$

is shifted to 70S formation. The association constant was found to be $K_{70\text{S}} = 5.3 \times 10^7\ \text{M}^{-1}$ [92]. Importantly, the location of the equilibrium is greatly affected by the individual and combined effects of initiation factor presence. IF2 was suggested to exist mostly complexed with GTP under in vivo conditions [96].

Association of Initiation Factors to 30S

The binding of initiation factors IF1, IF2, and IF3 to ribosomal subunit 30S appears to occur rapidly and in a random fashion (as reviewed by Gualerzi and Pon [93]; Fig. 12, and step (2) in Fig. 11). The net reaction for initiation factor binding to the 30S ribosomal subunit is given by:

$$30\text{S} + \text{IF1} + \text{IF2} \cdot \text{GTP} + \text{IF3} \rightleftharpoons \underbrace{30\text{S} \cdot \text{IF1} \cdot \text{IF2} \cdot \text{GTP} \cdot \text{IF3}}_{30\text{S}\cdot\text{IF}\cdot\text{GTP}} \,. \tag{41}$$

The effective formation of $30\text{S} \cdot \text{IF} \cdot \text{GTP}$ is crucial for the subsequent reaction steps of overall translation initiation. Although translation initiation may still proceed in the absence of several or all initiation factors, the rate of translation

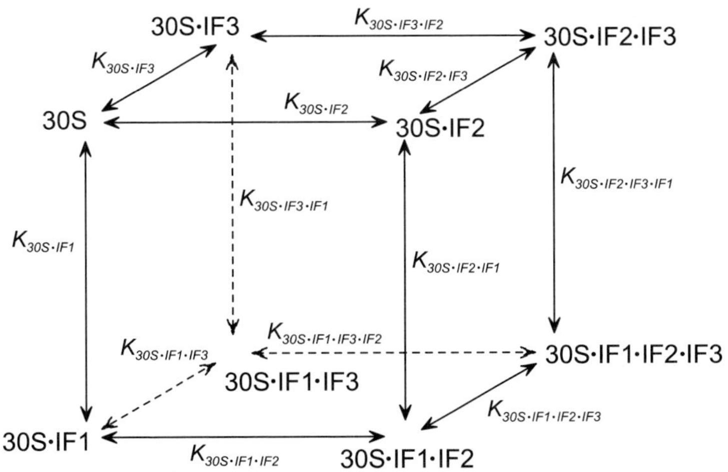

Fig. 12 Random order of binding of IF1, IF2, and IF3 to 30S. The preferred appearance of freely-dissolved IF2 in a complexed form with GTP is omitted in this representation

initiation is markedly enhanced only at sufficient levels of all three initiation factors [93, 95, 97].

An estimation of the various ribosomal complexes occurring during initiation site selection can be obtained from mass balancing and by using the corresponding association constants. The conservation relations for ribosomes and initiation factors are then obtained:

$$\begin{aligned} C_{30\mathrm{S},t} = {} & C_{30\mathrm{S}} + C_{30\mathrm{S}\cdot\mathrm{IF1}} + C_{30\mathrm{S}\cdot\mathrm{IF2}\cdot\mathrm{GTP}} + C_{30\mathrm{S}\cdot\mathrm{IF3}} \\ & + C_{30\mathrm{S}\cdot\mathrm{IF1}\cdot\mathrm{IF2}\cdot\mathrm{GTP}} + C_{30\mathrm{S}\cdot\mathrm{IF1}\cdot\mathrm{IF3}} + C_{30\mathrm{S}\cdot\mathrm{IF}} + C_{30\mathrm{S}\cdot\mathrm{IF2}\cdot\mathrm{GTP}\cdot\mathrm{IF3}} \\ & + C_{70\mathrm{S}} + \sum_{j=j_{\mathrm{R0}}}^{K} C_{70\mathrm{S},j} \end{aligned} \tag{42}$$

$$C_{50\mathrm{S},t} = C_{50\mathrm{S}} + C_{70\mathrm{S}} + \sum_{j=j_{\mathrm{R0}}}^{K} C_{70\mathrm{S},j} \tag{43}$$

$$C_{\mathrm{IF1},t} = C_{\mathrm{IF1}} + C_{30\mathrm{S}\cdot\mathrm{IF1}} + C_{30\mathrm{S}\cdot\mathrm{IF1}\cdot\mathrm{IF2}\cdot\mathrm{GTP}} + C_{30\mathrm{S}\cdot\mathrm{IF1}\cdot\mathrm{IF3}} + C_{30\mathrm{S}\cdot\mathrm{IF}} \tag{44}$$

$$\begin{aligned} C_{\mathrm{IF2},t} = {} & C_{\mathrm{IF2}\cdot\mathrm{GTP}} + C_{30\mathrm{S}\cdot\mathrm{IF2}\cdot\mathrm{GTP}} + C_{30\mathrm{S}\cdot\mathrm{IF}} + C_{30\mathrm{S}\cdot\mathrm{IF1}\cdot\mathrm{IF2}\cdot\mathrm{GTP}} \\ C_{\mathrm{IF2},t} = {} & + C_{30\mathrm{S}\cdot\mathrm{IF2}\cdot\mathrm{GTP}\cdot\mathrm{IF3}} \end{aligned} \tag{45}$$

$$C_{\mathrm{IF3},t} = C_{\mathrm{IF3}} + C_{30\mathrm{S}\cdot\mathrm{IF3}} + C_{30\mathrm{S}\cdot\mathrm{IF1}\cdot\mathrm{IF3}} + C_{30\mathrm{S}\cdot\mathrm{IF2}\cdot\mathrm{GTP}\cdot\mathrm{IF3}} + C_{30\mathrm{S}\cdot\mathrm{IF}} \,. \tag{46}$$

The summation term used in Eqs. 42 and 43 denotes the sum of ribosomes bound to mRNA (with K = number of base triplets within the coding region). Total concentrations of 30S and 50S ribosomal subunits are believed to exist in equal stoichiometric amounts in the reaction system. Initiation factor binding to 50S and 70S ribosomal subunits has been neglected owing to the reported low binding affinities [93, 98]. Substituting the association constants from Table 5 into Eqs. 42 to 46 leads to a set of nonlinear algebraic equations, which were then solved iteratively for the concentrations of uncomplexed species using OptdesX (Version 2.0.4, Design Synthesis, Inc.: Simulated annealing algorithm) and by minimizing the sum of squared relative errors. This procedure was also applied for computating the initial conditions to be used in dynamic simulations of protein production.

70S Initiation Complex Formation

The net reaction of 70S initiation complex formation (steps (3) to (6) in Fig. 11) comprises a multi-step mechanism, which was assumed to obey the scheme presented in Fig. 13. As can be viewed from this figure, a preinitiation complex is formed through the association of the ribosomal 30S subunit with initiator tRNA and the ribosome binding site (denoted by square brackets in step (1)).

Table 5 Association constants for computating levels of ribosomal complexes bound to initiation factors. Constants involving more than one initiation factor were derived using: 1.1×10^8 M^{-1} for IF1 binding to 30S in the presence of IF2 (Zucker and Hershey [92]), 3.6×10^7 M^{-1} for IF1 binding to 30S incubated with IF3 (Zucker and Hershey [92]), 1.2×10^8 M^{-1} for IF3 binding to 30S, when IF1 and IF2 were present (Chaires et al. [89]), 1.8×10^8 M^{-1} and 1.0×10^8 M^{-1} for the binding of IF2 and IF3, respectively, to 30S in the presence of both of the other initiation factors (Gualerzi and Pon [93]). 1 = Zucker and Hershey [92], 2 = Weiel and Hershey [90]

Parameter	Value	Source
K_{70S}	5.3×10^7 M^{-1}	1
$K_{30S \cdot IF1}$	5.0×10^5 M^{-1}	1
$K_{30S \cdot IF2}$	2.7×10^7 M^{-1}	2
$K_{30S \cdot IF3}$	3.1×10^7 M^{-1}	2
$K_{30S \cdot IF1 \cdot IF2 \cdot GTP}$	4.3×10^{14} M^{-2}	This study
$K_{30S \cdot IF1 \cdot IF3}$	5.6×10^{14} M^{-2}	dto.
$K_{30S \cdot IF2 \cdot GTP \cdot IF3}$	8.4×10^{14} M^{-2}	dto.
$K_{30S \cdot IF}$	3.7×10^{23} M^{-3}	dto.

Binding of fMet-tRNA$_f^M$ and the RBS, respectively, were assumed to be reversible and to take place randomly. A simplification inherently made is to consider the binding of either ligand to be unaffected by the binding of the other substrate. A slow rearrangment of this complex leads to the 30S initiation complex (30S-IC). The rate constant for this step, $k_{TLI,70SIC,1}$, was reported to be 0.1 s^{-1} [95].

$$30S{\cdot}IF{\cdot}GTP + fMet-tRNA_f^M + RBS \quad \underset{}{\overset{K_{M,fMet-tRNA_f^M} \cdot K_{RBS}}{\rightleftharpoons}} \quad \left[30S{\cdot}IF{\cdot}GTP{\cdot}fMet-tRNA_f^M{\cdot}RBS\right] \tag{1}$$

$$\left[30S{\cdot}IF{\cdot}GTP{\cdot}fMet-tRNA_f^M{\cdot}RBS\right] \quad \xrightarrow{k_{TLI,70SIC,1}} \quad \underbrace{30S{\cdot}IF{\cdot}GTP{\cdot}fMet-tRNA_f^M{\cdot}RBS}_{30S\text{-}IC} \tag{2}$$

$$30S-IC + 50S \quad \xrightarrow{k_{TLI,70SIC,2}} \quad 70S-IC + IF1 + IF3 \tag{3}$$

Fig. 13 Reaction steps involved in 70S initiation complex formation

Association of a 50S subparticle with the 30S initiation complex leads to the formation of the 70S initiation complex (70S-IC). During this reaction step, the positioning of fMet-tRNA$_f^M$ in the ribosomal P-site takes place together with a concomitant liberation of IF1 and IF3. (Rate constant $k_{TLI,70SIC,2}$ = 8.4×10^6 $M^{-1}s^{-1}$ was taken from Blumberg et al. [99]). The following rate expression was derived from Fig. 13 (Sect. B.1):

$$V_{\mathrm{TL1,70SIC}} = \frac{q_{j_{\mathrm{R0}}}^{\mathrm{R0}}\, V_{\mathrm{TLI,70SIC}}^{\max}}{D} \tag{47}$$

with

$$D = 1 + \frac{K_{\mathrm{M,fMet\text{-}tRNA}_f^{\mathrm{M}}}}{C_{\mathrm{fMet\text{-}tRNA}_f^{\mathrm{M}}}} + \frac{K_{\mathrm{M,RBS}}}{C_{\mathrm{RBS}}} + \frac{K_{\mathrm{M,50S}}}{C_{\mathrm{50S}}} + \frac{K_{\mathrm{M0,fMet\text{-}tRNA}_f^{\mathrm{M}}} K_{\mathrm{RBS}}}{C_{\mathrm{fMet\text{-}tRNA}_f^{\mathrm{M}}} C_{\mathrm{RBS}}} \,.$$

Parameter $q_{j_{\mathrm{R0}}}^{\mathrm{R0}}$ denotes the probability of the RBS being unoccupied (derived in Sect. 4). Other model parameters exhibit the following mathematical dependence on the rate constants and association constants of the elementary reactions:

$$V_{\mathrm{TLI,70SIC}}^{\max} = k_{\mathrm{TLI,70SIC,1}} C_{\mathrm{30S\cdot IF}} \tag{48}$$

$$K_{\mathrm{M,fMet\text{-}tRNA}_f^{\mathrm{M}}} = K_{\mathrm{fMet\text{-}tRNA}_f^{\mathrm{M}}} \tag{49}$$

$$K_{\mathrm{M,RBS}} = K_{\mathrm{RBS}} \tag{50}$$

$$K_{\mathrm{M,50S}} = \frac{k_{\mathrm{TLI,70SIC,1}}}{k_{\mathrm{TLI,70SIC,2}}} \,. \tag{51}$$

The affinity constants for initiator tRNA ($K_{\mathrm{M,fMet\text{-}tRNA}_f^{\mathrm{M}}}$) and mRNA ($K_{\mathrm{M,RBS}}$) were reported to be 0.05 μM and 0.009 μM, respectively [98, 100]. $K_{\mathrm{M,50S}} =$ 12 nM was calculated using the rate constants cited above. Throughout this study, the concentration of ribosome binding site (C_{RBS}) was taken to be equal to the concentration of the initiation codon ($C_{j_{\mathrm{R0}}}^{\mathrm{M}}$). In simulation analyses, Met-tRNA$_f^{\mathrm{M}}$ was supplied initially in sufficient amounts and then consumed over the course of the reaction.

IF2-Dependent GTP Hydrolysis

The ejection of IF2 from the 70S initiation complex (step (7) in Fig. 11) is accompanied by GTP hydrolysis due to

$$70\mathrm{S-IC} \xrightarrow{k_{\mathrm{TLI,IF2D}}} 70\mathrm{S}\cdot\mathrm{fMet-tRNA}_f^{\mathrm{M}}\cdot\mathrm{RBS} + \mathrm{IF2} + \mathrm{GDP} + \mathrm{Pi}\,. \tag{52}$$

This reaction was considered to follow first-order kinetics according to

$$V_{\mathrm{TLI,IF2D}} = k_{\mathrm{TLI,IF2D}} C_{\mathrm{70SIC}} \,. \tag{53}$$

The rate constants for IF2-dependent GTP hydrolysis and the release of inorganic phosphate were found to be 30 s^{-1} and 1.5 s^{-1}, respectively [96]. In the assumed mechanism, both reaction steps were combined into one step using a rate constant of 1.5 s^{-1}, in order to account for the slower of the reaction steps.

5.3 Elongation

Under physiological conditions, chain elongation proceeds at a rate of 10 to 20 aa/s [101]. The rate of elongation may be found to vary greatly along the mRNA [81, 84]. Elongation rate is kinetically influenced by (a) substrate availability (abundance of amino acids and tRNA [80]), modulated by (b) codon usage [102] and the strength of the codon-anticodon interaction [83], affected by (c) sterical hindrance between ribosomes travelling further downstream [86], and additionally regulated by (d) mRNA secondary structure [102, 103]. Furthermore, elongation factors catalyzing various steps of translation elongation are critically needed for maintaining high elongation rates. In the absence of elongation factors, the rate of protein synthesis is reduced by up to a factor of 10^4 [104].

5.3.1 Previous Modeling

The kinetics of GTP hydrolysis by EFG bound to ribosomes have been studied previously [105]. The formation rate of EFTu·GTP at EFTu regeneration was modeled kinetically and used for parameter estimation of substrate affinities [106]. The tRNA cycle was modeled in a probabilistic approach assigning mean duration times for various reaction steps [18]. Intricate kinetic models for tRNA charging have been developed to account for a functional dependency on Mg^{2+} ion concentration and the inhibitory influence of byproduct inorganic pyrophosphate [107, 108]. In modeling ternary complex formation between EFTu, GTP and aa-tRNA, a negative correlation of the abundance of aa-tRNA families and their affinities for EFTu·GTP was determined [102]. Pavlov and Ehrenberg [109] expressed the overall rate constant of elongation in terms of the total concentrations of EFTu and EFG.

A reaction scheme of the entire elongation cycle was proposed containing the regeneration of EFTu and EFG [110, 111]. Various ordered and random steady-state kinetic mechanisms were analyzed theoretically for both factorless and factor-dependent translation elongation [112, 113].

A matrix of translational efficiencies was derived in a statistical model [13]. The matrix elements denoted the efficiencies with which each aa-tRNA anticodon paired with a codon. In the same context, Solomovici et al. [118] computed elongation rates of synonymous codons given the hypothesis of an optimized (most economical) translation process.

Very detailed kinetic studies using stopped-flow techniques investigated elongation kinetics and identified rate constants for various steps of ligand association and catalytic isomerization [114].

5.3.2
Reaction Scheme and Kinetics

The subsequent model of translation elongation accounts for the processes of ternary complex formation, translation elongation, EFTu regeneration, and EFG regeneration.

Ternary Complex Formation

EFTu associates with GTP prior to formation of the ternary complex EFTu · GTP · aa-tRNA$_j$ (further on denoted by symbol T3$_j$ as well). The index j denotes any of the tRNA species. Free EFTu can bind with either GTP or GDP, according to

$$\mathrm{EFTu} + \mathrm{GTP} \underset{k_{-1}}{\overset{k_1}{\rightleftharpoons}} \mathrm{EFTu} \cdot \mathrm{GTP} \tag{54}$$

$$\mathrm{EFTu} + \mathrm{GDP} \underset{k_{-2}}{\overset{k_2}{\rightleftharpoons}} \mathrm{EFTu} \cdot \mathrm{GDP}\,. \tag{55}$$

The respective binding constant together with the rate constants for the elementary steps of association and dissociation were given by Romero et al. [116] for both GTP (8.0×10^6 M^{-1}, 2.0×10^5 M^{-1}s^{-1}, 2.5×10^{-2} s^{-1}) and GDP (5.3×10^8 M^{-1}, 9.0×10^5 M^{-1}s^{-1}, 1.7×10^{-3} s^{-1}), respectively.

The rate of ternary complex formation was derived for the forward and reverse reaction according to second-order kinetics on the basis of general collision theory [116]

$$V_{\mathrm{T3,Form},j} = k_{\mathrm{T3,Form},j} c_{\mathrm{EFTu \cdot GTP}} c_{\mathrm{aa\text{-}tRNA},j} - k_{-\mathrm{T3,Form},j} c_{\mathrm{T3},j}\,. \tag{56}$$

Rate constants for association and dissociation used in Eq. 56 may be discriminated against the type of aa-tRNA species. However, due to lack of information, they were taken in this study to be the same for each sort of aa-tRNA. The values applied were $k_{\mathrm{T3,Form}} = 5.0 \times 10^7$ M^{-1}s^{-1} and $k_{-\mathrm{T3,Form}} = 1$ s^{-1}, respectively, which were determined earlier for Trp-tRNA [110, 115]. Due to a relatively minor binding capacity [116], EFTu·GDP binding to aa-tRNA was omitted.

Translation Elongation

During an elongation cycle, the ribosome propagates from codon j to codon $j+1$ along the mRNA at the same time prolonging the nascent peptide chain by one amino acid and catalyzing the release of the tRNA of the previous elongation cycle according to

$$70S_j + \text{EFTu} \cdot \text{GTP} \cdot \text{aa-tRNA}_{j+1} + \text{EFG} \cdot \text{GTP} \xrightarrow{k_{\text{TLE},j}} 70S_{j+1} + \text{EFTu} \cdot \text{GDP} + \text{EFG} \cdot \text{GDP} + 2\text{Pi} + \text{tRNA}_j \,. \quad (57)$$

Translation factors EFTu and EFG occurring as various complexed species are treated as substrates and products of the overall reaction. The entire cycle can be divided into the reaction steps displayed in Fig. 14.

Symbol $70S_j$ denotes a ribosome which carries a peptide of j amino acids (P_j) that is attached to the tRNA in the ribosomal P-site (T_j^P). The association of ternary complex (aa-T_{j+1}·EFTu·GTP) takes place to a vacant ribosomal A-site (step (1) in Fig. 14). The act of ternary complex binding is reversible, which is of vital importance to correct tRNA selection and to proofreading. In a next step, the ribosome-bound ternary complex undergoes GTP hydrolysis (step (2)). Several conformational changes take place prior to EFTu·GDP release [124]. These isomerizations are summarized in reaction step (3). Through peptide bond formation, the growing polypeptide is prolonged by one amino acid (step (4)). During this step, the polypeptide chain attached to the tRNA in the P-site is handed over to the aa-tRNA located in the A-site. After this very rapid reaction step, a deacylated tRNA remains in the P-site. Binding of EFG·GTP (step (5)) is required to provide the energy needed for subsequent translocation. During translocation (step (6)), peptidyl-tRNA is transferred back into the P-site with the simultaneous release of the discharged tRNA (symbol T_j). This reaction is accompanied by GTP hydrolysis and by the propagation

$$(P_j - T_j^P) \cdot 70S_j + \text{aa} - T_{j+1} \cdot \text{EFTu} \cdot \text{GTP} \underset{k_{-1}}{\overset{k_1}{\rightleftharpoons}} (P_j - T_j^P) \cdot 70S_j \cdot \text{aa} - T_{j+1}^A \cdot \text{EFTu} \cdot \text{GTP} \quad (1)$$

$$(P_j - T_j^P) \cdot 70S_j \cdot \text{aa} - T_{j+1}^A \cdot \text{EFTu} \cdot \text{GTP} \xrightarrow{k_2} (P_j - T_j^P) \cdot 70S_j \cdot \text{aa} - T_{j+1}^A \cdot \text{EFTu} \cdot \text{GDP} + P_i \quad (2)$$

$$(P_j - T_j^P) \cdot 70S_j \cdot \text{aa} - T_{j+1}^A \cdot \text{EFTu} \cdot \text{GDP} \xrightarrow{k_3} (P_j - T_j^P) \cdot 70S_j \cdot \text{aa} - T_{j+1}^A + \text{EFTu} \cdot \text{GDP} \quad (3)$$

$$(P_j - T_j^P) \cdot 70S_j \cdot \text{aa} - T_{j+1}^A \xrightarrow{k_4} T_j^P \cdot 70S_j \cdot (P_{j+1} - T_{j+1}^A) \quad (4)$$

$$T_j^P \cdot 70S_j \cdot (P_{j+1} - T_{j+1}^A) + \text{EFG} \cdot \text{GTP} \underset{k_{-5}}{\overset{k_5}{\rightleftharpoons}} T_j^P \cdot 70S_j \cdot (P_{j+1} - T_{j+1}^A) \cdot \text{EFG} \cdot \text{GTP} \quad (5)$$

$$T_j^P \cdot 70S_j \cdot (P_{j+1} - T_{j+1}^A) \cdot \text{EFG} \cdot \text{GTP} \xrightarrow{k_6} (P_{j+1} - T_{j+1}^P) \cdot 70S_{j+1} \cdot \text{EFG} \cdot \text{GDP} + T_j + P_i \quad (6)$$

$$(P_{j+1} - T_{j+1}^P) \cdot 70S_{j+1} \cdot \text{EFG} \cdot \text{GDP} \xrightarrow{k_7} (P_{j+1} - T_{j+1}^P) \cdot 70S_{j+1} + \text{EFG} \cdot \text{GDP} \quad (7)$$

Fig. 14 Reaction steps involved in translation elongation cycle (as derived from Gast [110] and Pingoud et al. [115])

of the ribosome to the next codon on the mRNA. The dissociation of EFG·GDP (step (7)) completes the elongation cycle.

From the reaction scheme depicted in Fig. 14, and additionally considering the fact that codons can be recognized by more than one tRNA anticodon, steady state kinetics for the elongation cycle at codon j were derived using the symbolic computation (Sect. B.2):

$$V_{\mathrm{TLE},j} = \frac{q_j^{\mathrm{R}} V_{\mathrm{TLE},j}^{\max}}{1 + \frac{K_{\mathrm{M,T3}_j}}{\sum_i C_{\mathrm{T3}_j,i}} + \frac{K_{\mathrm{M,EFG\cdot GTP}}}{C_{\mathrm{EFG\cdot GTP}}}} . \tag{58}$$

The probability q_j^{R}, of codon $j + 1$ being unoccupied, was introduced earlier (Sect. 4). Other model parameters in Eq. 58 are composed from the rate constants for the elementary reaction steps (Fig. 14). Substituting the elementary rate constants provided by Gast [110], $K_{\mathrm{M,EFG}} \cdot \mathrm{GTP}$ results in a value of 0.22 μM. Total cellular contents of 44 tRNA species (out of the 46 tRNAs known to exist in *E. coli*) were provided by Dong et al. [117]. Parameter $K_{\mathrm{M,T}}3_j$ was selected to be equal to 0.4 μM.

The summation term depicted in Eq. 58 is the sum of ternary complexes with tRNA species that carry a correct amino acid corresponding to codon j and that are recognized by this codon. An example where the summation term comprises more than one element is codon UUG. This base triplet is matched by both $\mathrm{tRNA}^{\mathrm{Ser1}}$ and $\mathrm{tRNA}^{\mathrm{Ser5}}$ [117]. The rate of translation elongation at codon UUG is thus influenced by the concentrations of the respective ternary complexes corresponding to both of these tRNAs.

The maximum rate of translation elongation (symbol $V_{\mathrm{TLE},j}^{\max}$ in Eq. 58) is denoted by the concentration of ribosomes bound to codon j, and a codon-specific rate constant ($k_{\mathrm{TLE},j}$), according to

$$V_{\mathrm{TLE},j}^{\max} = k_{\mathrm{TLE},j}\, C_j^{\mathrm{R}} . \tag{59}$$

Codon-specificity may arise, for example, due to different binding strengths of codon-anticodon interaction for different tRNAs. The constant $k_{\mathrm{TLE},j}$ was calculated from

$$k_{\mathrm{TLE},j} = f_j\, k_{\mathrm{TLE}}^{\max} . \tag{60}$$

The efficiency factor, f_j, was adopted from Solomovici et al [118], who tabulated values of this parameter for all 61 sense codons. Unless otherwise stated, a maximum rate constant for translation elongation ($k_{\mathrm{TLE}}^{\max}$) of 24 codons/s was applied throughout this study.

In summary, the kinetic rate expression for translation elongation accounts for individual tRNA abundance of natural types of bacterial tRNA, codon-specific efficiency of translation elongation, steric interference among translating ribosomes, and the possibility of considering different affinities ($K_{\mathrm{M,T3}_j}$) for ternary complex selection at codon j.

EFTu Regeneration

Considering reversible ping-pong bi-bi kinetics (as suggested by Romero et al. [116]), the rate equation for the EFTu recycling can be derived to give

$$v_{\mathrm{EFTu-Reg}} = \frac{V_\mathrm{f}\left(C_\mathrm{A}\,C_\mathrm{B} - \frac{C_\mathrm{P}C_\mathrm{Q}}{K_{\mathrm{eq,EFTu}}}\right)}{D} \tag{61}$$

with

$$\begin{aligned} D =& K_{\mathrm{M,B}}C_\mathrm{A} + K_{\mathrm{M,A}}C_\mathrm{B} + \frac{V_\mathrm{f}}{V_\mathrm{r}K_{\mathrm{eq,EFTu}}}C_\mathrm{P}C_\mathrm{Q} + \frac{V_\mathrm{f}K_{\mathrm{M,P}}}{V_\mathrm{r}K_{\mathrm{eq,EFTu}}}C_\mathrm{Q} \\ &+ \frac{V_\mathrm{f}K_{\mathrm{M,Q}}}{V_\mathrm{r}K_{\mathrm{eq,EFTu}}}C_\mathrm{P} + C_\mathrm{A}\,C_\mathrm{B} + \frac{K_{\mathrm{M,A}}}{K_{i\mathrm{Q}}}C_\mathrm{B}C_\mathrm{Q} + \frac{V_\mathrm{f}K_{\mathrm{M,Q}}}{V_\mathrm{r}K_{\mathrm{eq,EFTu}}K_{i\mathrm{A}}}C_\mathrm{A}\,C_\mathrm{P}\,. \end{aligned}$$

Kinetic constants of Eq. 61 are listed in Table 6. The maximum forward rate is

Table 6 Kinetic constants of EFTu regeneration were calculated from the rate constants for the individual reaction steps given by Romero et al. [116] unless otherwise noted. Other parameter values were taken from [a]Ruusala et al. [119] and [b]Hwang and Miller [106]

		A EFTu·GDP	B GTP	P GDP	Q EFTu·GTP
K_M	(μM)	2.5[a]	50	3[b]	1
K_i	(μM)	5.6	6.5	15	1

$$V_\mathrm{f} = k_{\mathrm{EFTs,f}}C_{\mathrm{EFTs,t}}\,. \tag{62}$$

Symbol $C_{\mathrm{EFTs,t}}$ is the total concentration of EFTs. The maximum rate of the reverse reaction was calculated to be $V_\mathrm{r} = k_{\mathrm{EFTs,r}}C_{\mathrm{EFTs,t}}$. Constants $k_{\mathrm{EFTs,f}}$ and $k_{\mathrm{EFTs,r}}$ were reported to be 30 s^{-1} and 10 s^{-1}, respectively [119]. The equilibrium constant $K_{\mathrm{eq,EFTu}}$ was 0.19 using the rate constants published by Romero et al. [116].

EFG Regeneration

The regeneration of elongation factor EFG takes place spontaneously according to

$$\mathrm{EFG\cdot GDP} \underset{k_{-1}}{\overset{k_1}{\rightleftharpoons}} \mathrm{EFG + GDP} \tag{63}$$

$$\mathrm{EFG + GTP} \underset{k_{-2}}{\overset{k_2}{\rightleftharpoons}} \mathrm{EFG\cdot GTP}\,. \tag{64}$$

Values used for the association and dissociation rate constants of GDP binding were $2.7 \times 10^7\ \mathrm{M}^{-1}\mathrm{s}^{-1}$ and $100\ \mathrm{s}^{-1}$, respectively [110]. The rate constants for the forward and reverse reactions of Eq. 7 were reported to be $1.0 \times 10^7\ \mathrm{M}^{-1}\mathrm{s}^{-1}$ and $400\ \mathrm{s}^{-1}$, respectively [110].

Mass Conservation

Neglecting any uncomplexed EFTu, the total mass balance for elongation factors and involved guanylates can be represented by

$$C_{\mathrm{EFTu,t}} = C_{\mathrm{EFTu\cdot GTP}} + C_{\mathrm{EFTu\cdot GDP}} + \sum_{j=1}^{A} C_{\mathrm{T3},j} \tag{65}$$

$$C_{\mathrm{EFG,t}} = C_{\mathrm{EFG}} + C_{\mathrm{EFG\cdot GTP}} + C_{\mathrm{EFG\cdot GDP}} \tag{66}$$

$$C_{\mathrm{GTP,t}} = C_{\mathrm{GTP}} + C_{\mathrm{EFTu\cdot GTP}} + C_{\mathrm{EFG\cdot GTP}} \tag{67}$$

$$C_{\mathrm{GDP,t}} = C_{\mathrm{GDP}} + C_{\mathrm{EFTu\cdot GDP}} + C_{\mathrm{EFG\cdot GDP}} \,. \tag{68}$$

A is the number of different types of amino acids (usually 20). Elongation factor EFTs was regarded to function as a pure catalyst, whose concentration in the uncomplexed conformation is at any instant in time taken to be given approximately by the total concentration of this factor. Eqs. 65 to 68 were solved to yield the respective equilibrium concentrations of uncomplexed components together with their complexed counterparts.

5.4 Termination

The overall reaction stoichiometry considered for translation termination is given by

$$70\mathrm{S}_K + \mathrm{GTP} + \mathrm{H_2O} \xrightarrow{k_{\mathrm{TLT}}} 70\mathrm{S} + \mathrm{mRNA} + \mathrm{Protein} + \mathrm{tRNA}_K + \mathrm{GDP} + \mathrm{P}_i \,. \tag{69}$$

Release factors 1 (RF1) and 2 (RF2) assist in recognizing translational termination sites, which are signaled by the nonsense codons UAA, UAG, and UGA. Moreover, release factors RF3, RRF and RFH are known to be enrolled in translation termination [120]. These factors are, however, disregarded in this study, due to the limited information about their mechanistic involvement.

Allowing for a random order of substrate binding, and taking the reactions of substrate association to be rapid, the kinetic rate equation for translational termination can be derived as follows:

$$V_{\mathrm{TLT}} = \frac{V_{\mathrm{TLT}}^{\mathrm{max}}}{1 + \dfrac{K_{\mathrm{M,R}_K}}{C_K{}^{\mathrm{R}}} + \dfrac{K_{\mathrm{M,GTP}}}{C_{\mathrm{GTP}}} + \dfrac{K_{\mathrm{M,R}_K} K_{\mathrm{M,GTP}}}{C_K^{\mathrm{R}} C_{\mathrm{GTP}}}} \,. \tag{70}$$

The maximum termination rate $V_{\mathrm{TLT}}^{\max} = k_{\mathrm{TLT}} C_{\mathrm{RF}}$. Symbol C_{RF} represents the concentration of the proper release factor corresponding to the particular stop codon of the termination site. C_K^{R} is the concentration of ribosomes bound to codon K. The rate constants for termination were reported to be 0.25 s^{-1} for RF1, and 0.5 s^{-1} for RF2 [121]. The affinity constant of ribosomes with respect to RF1 was found to be $K_{\mathrm{M,RF1}} = 8.3$ nM [121]. Under the assumption that this parameter equals the dissociation rate constant, the same value was taken for parameter $K_{\mathrm{M,R}_K}$. The constant $K_{\mathrm{M,GTP}}$ was selected to be equal to 20 μM.

5.5 tRNA Charging

The charging of tRNA with amino acids is promoted by the aminoacyl-tRNA-synthetases (ARS), thereby consuming ATP and releasing AMP and inorganic pyrophosphate. The net stoichiometry reads

$$\mathrm{aa} + \mathrm{tRNA} + \mathrm{ATP} \xrightarrow{\mathrm{ARS}} \mathrm{aa\text{-}tRNA} + \mathrm{AMP} + \mathrm{PP}_i \,. \tag{71}$$

For each amino acid, there exists at least one corresponding ARS [122]. Assuming a rapid equilibrium binding of substrates and neglecting product inhibition terms, the following rate equation was considered to apply for the reaction of tRNA charging:

$$V_{\mathrm{ARS},i,k} = \frac{V_{\mathrm{ARS},i,k}^{\max}}{D} \tag{72}$$

with

$$D = 1 + \frac{K_{\mathrm{M,ARS,aa}_j}}{C_{\mathrm{aa},j}} + \frac{K_{\mathrm{M,ARS,ATP}}}{C_{\mathrm{ATP}}} + \frac{K_{\mathrm{M,ARS,tRNA}_j}}{C_{\mathrm{tRNA}_j}} \,.$$

In analogy to parameter values given by Hirshfield and Yeh [123], $K_{\mathrm{M,ARS,aa}_j}$ and $K_{\mathrm{M,ARS,ATP}}$ were considered to be equal to 20 μM and 100 μM, respectively. Constants $K_{\mathrm{M,ARS,tRNA}_j}$ and k_{cat} were adopted from Schulman and Pelka [124] and Schulman [125], and were 0.5 μM and 1.0 s^{-1}, respectively. In a simplifying assumption, the kinetic constants displayed in Eq. 72 were taken to be the same for all tRNA species, and for all aa-tRNA synthetases.

The formylation reaction of methionine bound to initiator tRNA was disregarded in this study. In simulation analyses, fMet-$\mathrm{tRNA}^{\mathrm{M}_f}$ was supplied initially in sufficient amounts and then consumed over the course of the process.

5.6 Model Reduction

Applying the model simplification of merging groups of codons, as suggested earlier (Sect. 4), causes a profound effect on material balancing of variables enrolled in the translation process. In this case, the rate of translation elongation condenses multiple (say n_c) elongation cycles together. The reaction stoichiometry then reads:

$$70S'_j + \sum_{k=1}^{n_c} \text{EFTu} \cdot \text{GTP} \cdot \text{aa-tRNA}_{j+1,k} + n_c\text{EFG} \cdot \text{GTP40} \tag{73}$$

$$\overset{k_{\text{TLE},j}}{\rightarrow} 70S'_{j+1} + n_c\text{EFTu} \cdot \text{GDP} + 2n_c\text{P}_i + n_c\text{EFG} \cdot \text{GDP} + \sum_{k=1}^{n_c} \text{tRNA}_{j,k} \,.$$

Combining multiple rounds of the reaction scheme given in Fig. 14, it can be shown (see Sect. B.2) that the overall kinetics of n_c elongation steps may be described mathematically by

$$V'_{\text{TLE},j} = \frac{q_j^{\text{R}} k'_{\text{TLEj}} \, C_j^{\text{R}'}}{1 + \sum\limits_{k=1}^{n_c} \frac{K_{\text{M,T3}_j}}{\sum\limits_i C_{\text{T3}_j,i,k}} + \frac{K_{\text{M,EFG}\cdot\text{GTP}}}{C_{\text{EFG}\cdot\text{GTP}}}} \,. \tag{74}$$

The prime refers to state variables of the new codon grid, with each position j reflecting n_c codons at once. In an approximation, parameter $k'_{\text{TLE},j}$ was calculated from the smallest of the efficiency factors within each group of n_c codons in the reduced state representation, according to

$$k'_{\text{TLE},j} = \min(f_{j,k}) \frac{k_{\text{TLE}}^{\max}}{n_c} \quad \text{with} \quad k = 1 \text{ to } n_c \,. \tag{75}$$

The sum of elongations consuming a particular ternary complex k is given by

$$V_{\text{SumT3},k} = \sum_{j=j_{\text{R0}}}^{K-1} \alpha_{j,k} V_{\text{TLE},j} \,. \tag{76}$$

Parameter $\alpha_{j,k}$ denotes the fraction of translational elongation rates j at which the kth ternary complex is consumed. $\alpha_{j,k}$ typically equals 1 when only one cognate ternary complex exists. $\alpha_{j,k}$ takes values between 0 and 1 when codons are matched by more than one tRNA. $\alpha_{j,k}$ equals 0 for codons j that do not relate to the kth tRNA. This parameter was subsequently approximated by the ratio of the total concentration of the kth ternary complex involved in elongation at a particular codon j to the sum of the total concentrations of ternary complexes recognized by this codon. That is,

$$\alpha_{j,k} \approx \frac{C_{\text{T3},j,k}}{\sum\limits_i C_{\text{T3},j,i}} \quad \text{for} \quad j_{\text{R0}} \le j \le K-1 \,. \tag{77}$$

Analogously to Eq. 76, the sum of elongation rates releasing an uncharged tRNA species k may be written as

$$V_{\mathrm{SumT},k} = \sum_{j=j_{\mathrm{R0}+1}}^{K} \alpha_{j,k} V_{\mathrm{TLE},j} \, . \tag{78}$$

5.7 Material Balances

The following material balances cover the time-dependent changes in protein product, concentrations of ribosomes freely dissolved and in diverse states of complexation with translation factors, as well as when they are bound to mRNA in different positions. Material balancing further includes balances for the full sets of amino acids (aa $_i$), tRNA species (T_k), aminoacylated tRNAs, ternary complexes EFTu·GTP·aa-tRNA $_j$ ($T3_k$), and balances of energy components consumed during translation.

$$\frac{dC_{\mathrm{Protein}}}{dt} = V_{\mathrm{TLT}} \tag{79}$$

$$\frac{dC^{\mathrm{R}^*}_{j\mathrm{R0}}}{dt} = V_{\mathrm{TLI,70SIC}} - V_{\mathrm{TLI,IF2D}} \tag{80}$$

$$\frac{dC^{\mathrm{R}}_{j\mathrm{R0}}}{dt} = V_{\mathrm{TLI,IF2D}} - V_{\mathrm{TLE},j_{\mathrm{R0}}} \tag{81}$$

$$\frac{dC^{\mathrm{R}}_{j}}{dt} = V_{\mathrm{TLE},j-1} - V_{\mathrm{TLE},j} \quad \text{for} \quad j_{\mathrm{R0}} \le j \le k \tag{82}$$

$$\frac{dC^{\mathrm{R}}_{K}}{dt} = V_{\mathrm{TLE},K-1} - V_{\mathrm{TLT}} \tag{83}$$

$$\frac{dC_{\mathrm{aa}_i}}{dt} =- \sum_{k=1}^{T} V_{\mathrm{ARS},i,k} \quad \text{for} \quad 1 \le i \le A \tag{84}$$

$$\frac{dC_{T_k}}{dt} = V_{\mathrm{Sum}T,k} - V_{\mathrm{ARS},i,k} \quad \text{for} \quad 1 \le k \le T \tag{85}$$

$$\frac{dC_{\mathrm{fMet-tRNA}_f^{\mathrm{M}}}}{dt} =- V_{\mathrm{TLI,70SIC}} \tag{86}$$

$$\frac{dC_{\mathrm{aa}_i\text{-}\mathrm{TRNA}_k}}{dt} = V_{\mathrm{ARS},i,k} - V_{\mathrm{T3Form},k} \quad \text{for} \quad 1 \le k \le T \tag{87}$$

$$\frac{dC_{\mathrm{T3}_k}}{d} = V_{\mathrm{T3Form},k} - V_{\mathrm{SumT3},k} \quad \text{for} \quad 1 \le k \le T \tag{88}$$

$$\frac{dC_{\mathrm{ATP}}}{dt} =- \sum_{i=1}^{A} \sum_{k=1}^{T} V_{\mathrm{ARS},i,k} \tag{89}$$

$$\frac{dC_{\mathrm{AMP}}}{dt} = \sum_{i=1}^{A} \sum_{k=1}^{T} V_{\mathrm{ARS},i,k} \tag{90}$$

$$\frac{dC_{\mathrm{GTP}}}{dt} = - V_{\mathrm{TLI,IF2D}} - V_{\mathrm{EFTu\text{-}Reg}} - V_{\mathrm{TLT}} - V_{\mathrm{EFG\text{-}GTP,Ass}} \tag{91}$$

$$\frac{dC_{\mathrm{GDP}}}{dt} = V_{\mathrm{TLI,IF2D}} + V_{\mathrm{EFTu\text{-}Reg}} + V_{\mathrm{TLT}} - V_{\mathrm{EFG\cdot GDP,Ass}} \tag{92}$$

$$\frac{dC_{\mathrm{EFG\text{-}GTP}}}{dt} = V_{\mathrm{EFG\text{-}GTP,Ass}} - \sum_{k=1}^{T} V_{\mathrm{SumT3},k} \tag{93}$$

$$\frac{dC_{\mathrm{EFG\text{-}GDP}}}{dt} = \sum_{k=1}^{TV_{\mathrm{Sum\ T3},k}} V_{\mathrm{SumT3},k} + V_{\mathrm{EFG\text{-}GDP,Ass}} \tag{94}$$

$$\frac{dC_{\mathrm{EFTu\text{-}GTP}}}{dt} = V_{\mathrm{EFTu\text{-}Reg}} - \sum_{j=j_{\mathrm{R0}}}^{K-1} \sum_{k=1}^{T} V_{\mathrm{T3Form},k} \tag{95}$$

$$\frac{dC_{\mathrm{EFTu\text{-}GDP}}}{dt} = \sum_{j=j_{\mathrm{R0}+1}}^{K} \sum_{k=1}^{T} V_{\mathrm{SumT3},k} - V_{\mathrm{EFTu\text{-}Reg}} \, . \tag{96}$$

Because functionality of the translation system relies on the combination of the different modules (transcription, degradation and translation) it is part of the strategy to miss out the isolated simulation of an "autonomous" translation module missing the emerging, non-additive effects. Instead, dynamic simulations of the translation module will be shown in the following section in context with the application of the aggregated model (transcription, degradation, translation) to the study of mutual interactions and combined effects of the various compounds within the example of cell-free protein expression. This system also serves as an experimental basis for validation of the integrated model.

6 Application to Cell-Free Protein Biosynthesis

6.1 Introduction

Cell-free protein synthesis systems are ideal, simplified exploration tools for gene expression analysis. Their main advantages arise from their reduced complexity in comparison to a growing organism and their convenient accessibility. In these in vitro systems, protein production is typically achieved on the basis of cellular lysates, which contain the required biocatalysts extracted from the living cell. By choosing substrate composition appropriately, it is possible to selectively activate the endogenous gene expression pathway, whereas the majority of regulatory mechanisms, for instance induction and repression encountered in vivo, are switched off. By employing recombinant DNA technology, the synthesis capacity and energy expenditures usually spent on cell growth can thus in principle be redirected towards the production of a single or a few gene products.

Cytotoxic and novel peptides following from the incorporation of unnatural amino acids, that are not expressed in vivo, have been synthesized in mg amounts in these cell extracts [126]. Practical examples of cell-free protein expression methods cover their use in functional genomics and evolutionary studies, such as in ribosomal display [127].

Although in vitro protein production has been used for several decades now, many of the original constraints limiting both production rates and process duration remain unresolved. While various modifications have been made to improve commonly-used systems [128, 129], for example by applying condensed extracts [130] and continuous substrate supplementation via dialysis membrane technology, the problem of poor volumetric productivities still exists. Typical volumetric protein synthesis rates achieved in *E. coli* cell extracts are about 0.5 mg/ml/h [131–133]. This value is roughly 300-fold lower than the in vivo synthesis rate of total protein at a specific growth rate of $\mu = 1.0\,h^{-1}$, calculated from Bremer and Dennis [101]. The particular causes of this discrepancy between in vitro and in vivo synthesis rates are unclear.

Although cell-free protein synthesis systems provide meaningful ways to probe gene expression models, they differ in some important aspects from the in vivo situation. For balanced growth, gene expression settles into a steady state, which is characterized by static pool concentrations and a constant renewal of the involved biocatalysts. On the other hand, cell-free gene expression systems suffer from a continuous catabolysis of supplied substrates and a gradual loss of biocatalytic activity. Countermeasures to this commonly include the use of an energy regeneration system, as well as the addition of protease and RNase inhibitors. Nevertheless, degradation processes affecting the translation apparatus cannot be completely ruled out. At the same time, the initial lysate composition, in terms of absolute and relative concentrations of translational key players, is altered in comparison to in vivo conditions. This is caused mainly by the various processing steps and dilutions applied during lysate production, which typically add up to an approximately 20-fold dilution in comparison to the living cell, as well as due to the supplementation of selected components such as translation factors and tRNA. Apart from sequence-specific gene expression kinetics, a mathematical description of in vitro protein biosynthesis therefore needs to take into account all of the in vitro specific properties as well.

In spite of its simplicity compared to in vivo conditions, modeling cell-free protein biosynthesis requires the formulation of the comprehensive gene expression model. An important issue is the emergent properties of the system

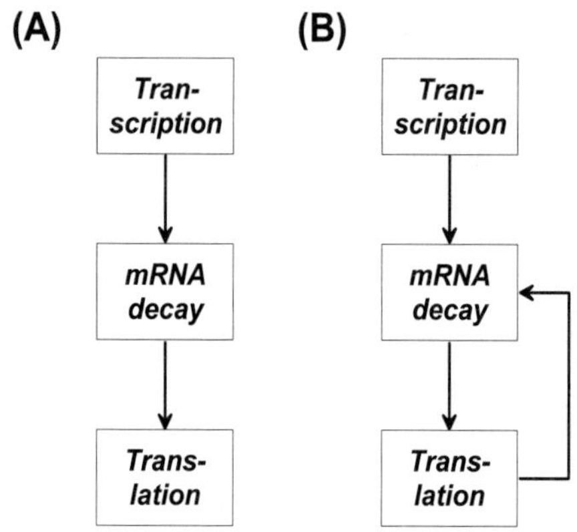

Fig. 15 Coupling of modeling tools (**a**) Unidirectional information flow (**b**) Feedback interaction

caused by the aggregation of the individual modules. This is schematically demonstrated in Figure 15. The sequential scheme displayed on the left hand side of this figure constitutes a picture of reality that is oversimplified. When coupling the modeling units of gene expression, non-additive effects also arise. An example of the nonlinearity of modular interactions is the feedback regulation of translational fidelity affecting mRNA degradation rate (see the right hand side of Fig. 15). Translating ribosomes are capable of providing a barrier to RNases trying to access endonucleolytic cleavage sites (Sect. 3 and Sect. A). In order to account for these phenomena in a gene expression system, it is necessary to adequately modify the stand-alone modeling units defined earlier.

In the following, we present the model adjustments that need to be made in order to arrive at a combined gene expression model. Moreover, the effects of energy regeneration, lysate composition, and inactivation kinetics – additional problems in the cell-free protein biosynthesis – are outlined. For the purpose of model verification, the augmented model is subsequently applied to simulate the performance of cell-free protein expression. Such an approach aims to explore the predictability of the model by comparing simulation results with experimentally-observed gene expression behavior.

6.2 Modeling and Simulation Tools

6.2.1 Combined Gene Expression Model

The mRNA synthesis rate for each base triplet j can be acquired by considering uniformly distributed RNA polymerases along the coding region. The time delay between initiation of transcript synthesis and the time point, when a particular base triplet j is synthesized, is neglected in this analysis. Due to the high specific transcription rate of T7 RNA polymerase, of about 100 to 250 nucleotides per second [134], both 5′ and 3′ transcript ends of mRNA were taken to be synthesized approximately simultaneously and at the same rate.

Since the processes of transcription and translation are highly energy-dependent, all aspects of protein synthesis need to be viewed within the context of energy recycling systems. Energy regeneration performs the task of continuously restoring the pools of energy-carriers (such as ATP and GTP) as they are constantly depleted over the course of protein synthesis. While these processes are maintained in the living cell as a result of catabolism, phosphor donors need to be added specifically to cell-free systems to spur these processes on. In addition, it is also necessary to supply the enzymes needed for regeneration, unless the regeneration machinery relies solely on endogenous enzymes that are already present in the native cellular extract.

6.2.2 Energy Regeneration

The enzyme acetate kinase reversibly catalyzes the phosphorylation of ADP to form ATP, while acetyl phosphate (AcP) is converted to acetate (Ac). A kinetic rate expression for *E. coli* acetate kinase was derived in this study from the data given by Janson and Cleland [135]. The kinetics are assumed to obey a rapid equilibrium random bi bi mechanism with additional formation of dead-end inhibition complexes EBQ (= E · AcP · ATP) and EBP (= E · AcP · Ac) according to

$$V_{\mathrm{Ack}} = \frac{V^{\max}_{\mathrm{Ack},f} V^{\max}_{\mathrm{Ack},r} \left(C_{\mathrm{ADP}} C_{\mathrm{AcP}} - \frac{C_{\mathrm{ATP}} C_{\mathrm{Ac}}}{K_{\mathrm{eq}}} \right)}{D} \tag{97}$$

with

$$\begin{aligned} D = {} & V^{\max}_{\mathrm{Ack},r} \left(K_{i,\mathrm{ADP}}\, K_{\mathrm{M,AcP}} + K_{\mathrm{M,AcP}} C_{\mathrm{ADP}} + K_{\mathrm{M,ADP}} C_{\mathrm{eAcP}} + C_{\mathrm{ADP}} C_{\mathrm{AcP}} \right) \\ & + \frac{V^{\max}_{\mathrm{Ack},f}}{K_{\mathrm{eq}}} \left[C_{\mathrm{Ac}} C_{\mathrm{ATP}} + \left(K_{\mathrm{M,ATP}} C_{\mathrm{Ac}} + K_{\mathrm{M,Ac}} C_{\mathrm{ATP}} \right) \left(1 + \frac{C_{\mathrm{AcP}}}{K_{i,\mathrm{AcP}}} \right) \right] . \end{aligned}$$

The enzyme adenylate kinase (Adk) performs the reaction, converting AMP and ATP into two molecules of ADP. The following reversible rate equation was assumed to be representative of the reaction

$$V_{\mathrm{Adk}} = \frac{V^{\max}_{\mathrm{Adk},f} C_{\mathrm{AMP}} C_{\mathrm{ATP}}}{\left(K_{\mathrm{M,AMP}} + C_{\mathrm{AMP}} \right) \left(K_{\mathrm{M,ATP}} + C_{\mathrm{ATP}} \right)} - \frac{V^{\max}_{\mathrm{Adk},r} \left(C_{\mathrm{ADP}} \right)^2}{\left(K_{\mathrm{M,ADP}} + C_{\mathrm{ADP}} \right)^2} . \tag{98}$$

Parameter values for model constants used in Eqs. 97 and 98 are listed in the Appendix (Sect. C.1). Apart from this enzyme, further nucleoside monophosphate kinases (Nmk) exist in *E.coli* to perform the reaction

$$N_1\mathrm{DP} + N_2\mathrm{TP} \longleftrightarrow N_1\mathrm{DP} + N_2\mathrm{DP}\,. \tag{99}$$

Nucleoside diphosphate kinase (Ndk) catalyzes the reaction

$$N_1\mathrm{DP} + N_2\mathrm{TP} \longleftrightarrow N_1\mathrm{TP} + N_2\mathrm{DP}\,. \tag{100}$$

Enzymes Ndk and Nmk form a network of near-equilibrium reactions, with both enzyme types exhibiting equilibrium constants close to unity [136]. Thus, and in order to mathematically implement the ability to regenerate each of the four ribonucleoside mono-and diphosphates, respectively, modeling assumed that three further enzymes exist that are analogous to acetate kinase and that are capable of regenerating nucleotides CDP, GDP, and UDP, respectively. By the same reasoning, rate expressions were also derived for three putative enzymes that were assumed to perform a reaction similar to the adenylate kinase reaction, except that they replace AMP with one of the nucleoside monophophates CMP, GMP, and UMP, respectively. Moreover, non-enzymatic chemical hydolysis of acetyl phosphate [137] was taken into account by a first-order decay reaction

$$V_{\mathrm{d,AcP}} = k_{\mathrm{d,AcP}} C_{\mathrm{AcP}}\,. \tag{101}$$

Endogenous nuclease activity hydrolyzing nucleoside triphosphates was accounted for with

$$V_{\mathrm{d,ATP}} = k_{\mathrm{d,ATP}} C_{\mathrm{ATP}}\,. \tag{102}$$

Analogous kinetic rate expressions were also derived for the hydrolysis of CTP, GTP, and UTP, respectively.

6.2.3 Catalyst Inactivation

Catalyst inactivation takes place inherently in cell-free protein synthesis systems. In particular, a significant reduction of ribosomal protein S1 was observed experimentally in proteome analysis by Schindler et al. [138], and has thus been accounted for in the modeling scheme. The inactivation of ribosomal protein S1 (RP-S1) was included in the model in terms of a first-order inactivation of the maximum rate of 70S initiation complex formation:

$$V^{\max}_{\mathrm{TLI,70SIC}} = k_{\mathrm{TLI,70SIC,1}}\, \mathrm{e}^{(-k_{\mathrm{d,RP\text{-}S1}} t)} C_{\mathrm{30S\cdot IF}}\,. \tag{103}$$

The time-dependent decrease of both EFTu and EFTs was modeled as a first-order decay affecting their respective total concentrations, according to

$$C_{\mathrm{EFTu},t} = C_{\mathrm{EFTu},t}(t=0)\, \mathrm{e}^{(-k_{\mathrm{d,EFTu}} t)} \tag{104}$$

$$C_{\mathrm{EFTs},t} = C_{\mathrm{EFTs},t}(t=0)\, \mathrm{e}^{(-k_{\mathrm{d,EFTs}} t)}\,. \tag{105}$$

Table 7 Half-life times of selected translational coponents calculated from experimental data [138]. RP-S1 = ribosomal protein S1, EFTu and EFTs are the elongation factors Tu and Ts respectively

Component	half-life [min]	k_D [1/min]
PP-S1	13	0.05382
EF-Tu	51	0.01364
EF-Ts	59	0.01166

The first-order degradation constants used in the above equations (Eq. 103 to Eq. 105) were calculated from experimental data [138] and are summarized in Table 7. These parameters were then substituted into the respective material balance equations derived earlier (Sect. 5).

In addition, the same inactivation of protein T7 RNA polymerase as identified for the isolated enzyme [44] was assumed to also apply to conditions of simultaneous transcription and translation. It remains unclear whether this assumption is also valid in cell-free protein synthesis systems, because the experimental conditions of both systems may not be comparable, for example with respect to total ion concentration and total protein concentration.

6.3 Materials and Methods

6.3.1 Plasmids

Plasmid pIVEX-2.1-GFP, coding for recombinant GFPuv, which is controlled by both T7-promoter and T7-terminator, was a kind gift from Roche Molecular Diagnostics, Germany. The molecular size of the plasmid was 4355 bp, the total length of the GFP-coding mRNA was 1041 bases. Plasmids used for in vitro studies were purified using the Qiagen Plasmid Maxi-Kit (Qiagen, Hilden, Germany).

6.3.2 Preparation of Cell-Free Crude Extract

Preparation of the S30-cell extract from *E. coli* A19 was performed according to Pratt [129] with modifications described previously [139]. The protein concentration of the final lysate was 29.5 mg/l, as measured by the Bradford assay (BioRad, Munich, Germany). The ribosome concentration was 7.5 μM, which was estimated from adsorption units AU260 nm of 290 according to Geigenmüller and Nierhaus [140]. For this purpose, 100 μl of the S30-lysate was diluted into 100 ml of bidistilled water. The adsorption of 1 ml of the 1 : 1000 diluted solution was measured at 260 nm. One adsorption unit per ml equals to 24 pmol of S70 ribosomes. Further, the ribosome concentration was additionally quantified by denaturing polyacryamide gels (5%) according to Sambrook et al. [142]. 10 μl of the lysate was diluted with 240 μl of 1% SDS. Afterwards, the total RNA was extracted by repeated phenol/chloroform extraction. Staining of the gel was performed with toluidene blue. Quantification was densiometrically performed using Pharmacia's ImageMaster software package and using the 16*S*/32*S* rRNA-calibration standard of known concentration (Roche Molecular Diagnostics, Germany). A total ribosome concentration of 12 μM was determined with respect to this quantification standard (100 A260 units; each of 0.1 μg/ml).

6.3.3 Coupled In Vitro Transcription/Translation

Coupled cell-free protein biosynthesis was performed using an S30 bacterial cell extract system generated from *E. coli* A19 according to Pratt [129], with minor modifications as previously described [139]. Batch-wise cell-free transcription/translation was performed at 30 °C and the reaction mixture contained the following components: The respective plasmid at a final concentration of 5.6 nM, 2 kU ml^{-1} T7-RNA polymerase, 48 mg ml^{-1} (m v-1) E. coli-tRNA, 100 mM Hepes/KOH, pH 7.6, 2 mM ATP, 1.6 mM GTP, 1 mM CTP, 1 mM UTP, 250 μM of all 20 amino acids, 18.8 μM folinic acid, 1 mg l^{-1} (m v-1) rifampicin, 100 mM KOAc, 18 mM Mg(OAc)2, 1 mM EDTA, 2 mM dithiothreitol, 0.03% (m v-1) sodium azide, and *E. coli* S30 extract at a final protein concentration of 5.9 g l^{-1} (m v-1) (equal to 1.5 μM total ribosome concentration). 40 mM acetyl phosphate and endogenous acetate kinase were used as an energy regeneration system.

6.3.4 Quantification of Protein Synthesized In Vitro

In vitro synthesized protein was estimated from the incorporation of radiolabeled ^{14}C-leucine: 66.7 μM of ^{14}C-leucine (11.7 GBq $mmol^{-1}$, Amersham Pharmacia Biotech, UK) was added to the standard mixture. At respective times, 4 μL aliquots were withdrawn and the concentration of the protein determined by liquid scintillation counting as described previously [44]. Aliquots of the reaction mixture were further analyzed by SDS-PAGE followed by autoradiography according to Katanaev et al. [141].

6.3.5
Measurements of Metabolites

Ionic Pair Chromatography on Reversed Phase RP18-column (GROM-SIL, GROM, Herrenberg, Germany/SpectraPhysics, San Jose, CA) was used with minor modifications according to Mailinger et al. [143] for measurements of all nucleotide concentrations (NXP). 30 μl of the reaction mixture were pipetted into 120 μl of hot (95 °C) 0.2 vol % phosphoric acid. After centrifugation, 100 μl of the clear supernatant was used for HPLC analysis. The concentration of acetyl phosphate was determined according to Lippmann and Tuttle [144]. In order to prevent spontaneous chemical hydrolysis, all reactions were handled on ice.

6.3.6
Measurement of mRNA Concentration

Total mRNA synthesized in the coupled system was estimated from incorporation of ^{14}C-ATP as described previously [44]. 200 μM of ^{14}C-ATP (1.92 GBq/mmol; Amersham Pharmacia Biotech, UK) was added to the standard mixture. At respective times, aliquots of 20 μl were taken, and the concentration (μM) of synthesized mRNA was estimated from the liquid scintillation assay as published by Arnold et al. [44]. The quality of synthesized mRNA was further analyzed on denaturing polyacrylamide gels (5% PAGE, 6 M urea) as described in the original study.

6.4
Dynamic Simulation

Figures 16 to 21 show the simulated time traces of selected quantities (mostly concentrations and reaction rates) characterizing cell-free synthesis of green fluorescent protein (GFP)under batch conditions. The model applied combines reactions involved in (a) mRNA synthesis, (b) mRNA degradation, (c) ribosomal translation, (d) energy regeneration, and (e) inactivation kinetics of proteins S1, EFTu, EFTs, and T7 RNA polymerase. For those components where measurements were made, simulation results are compared to their experimentally-determined counterparts.

The primary intention of this analysis was to investigate the predictive power of the model in comparison to experimental data. Due to the number of states and parameters contained in the model, and the uncertainty associated with model constants taken from the literature, the ability to qualitatively predict measured results was of greater concern to the analysis, rather than a quantitative description of system behavior. No particular parameter estimation procedure was performed here. Initial conditions for balanced concentrations are given in Table 8. These were obtained by considering a 20-fold dilution of proteins and ribosomes in cell-free systems in comparison to a growing *E. coli* cell [101].

As can be seen from Fig. 16, the predicted time dependencies of concentrations of protein GFP, full-length mRNA, and acetyl phosphate correspond quite favorably with the experimental observed dependencies. The concentrations of GFP and mRNA increase with time as they are synthesized. Protein concentration is seen to level off after about one hour into the experiment. This is primarily a consequence of the measured inactivation of ribosomal protein S1, with a half-life of 13 min (Table 7). The concentration of acetyl phosphate is seen to continuously diminish with time, mainly due to acetyl phosphate consumption through the acetate kinase reaction and its equivalents.

Due to energy regeneration, it is possible to maintain sufficiently high levels of nucleotide concentrations. This is demonstrated in Fig. 16c, where the time courses of the concentrations of adenylates and GTP are displayed. In con-

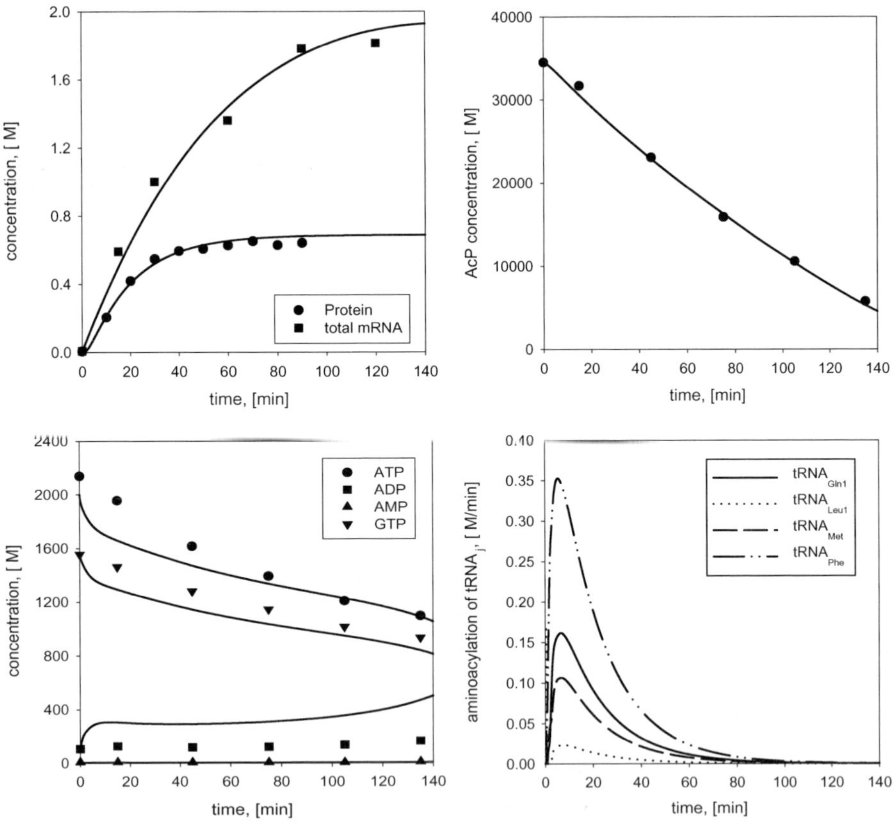

Fig. 16 Time courses of measured and predicted levels of (**a**) protein GFP and full-length mRNA, (**b**) acetyl phosphate, (**c**) ATP, ADP, AMP, and GTP, and (**d**) predicted rates of aminoacylation for selected tRNAs

Table 8 Various initial conditions used when simulating cell-free protein synthesis during optimization. Reference condition refers to the simulation study of Sect. 4. A - 30-fold EFTu concentration in comparison to the reference state. B - All EF concentrations raised by a factor of 30. C - Elevated IF levels. D - Simultaneous increase in the concentrations of both initiation factors and elongation factors

Concentration (μM)	Reference	A	B	C	D
$C_{30S_{tot}}$	1.40	1.40	1.40	1.40	1.40
$C_{EFG_{tot}}$	1.21	1.21	36.4	1.21	36.4
$C_{EFTu_{tot}}$	1.06	31.8	31.8	1.06	31.8
$C_{EFTs_{tot}}$	0.27	0.27	8.18	0.27	8.18
$C_{IF1_{tot}}$	0.38	0.38	0.38	1.67	1.67
$C_{IF2_{tot}}$	0.45	0.45	0.45	1.28	1.28
$C_{IF3_{tot}}$	0.30	0.30	0.30	1.65	1.65
C_{30S}	0.007	0.007	0.007	0.003	0.003
C_{50S}	0.32	0.32	0.32	1.24	1.24
C_{IF1}	0.07	0.07	0.07	0.45	0.45
C_{IF2}	0.11	0.11	0.11	0.07	0.07
C_{IF3}	0.01	0.01	0.01	0.42	0.42
C_{EFG}	0.02	0.02	0.66	0.02	0.66
$C_{EFG \cdot GTP}$	0.78	0.78	24.8	0.78	24.8
$C_{EFG \cdot GDP}$	0.41	0.41	10.9	0.41	10.9
$C_{EFTu \cdot GTP}$	0.71	25.6	26.2	0.71	26.2
$C_{EFTu \cdot GDP}$	0.35	6.26	5.62	0.35	5.62
C_{GTP}	1549	1530	1505	1549	1505
C_{GDP}	75.2	72.2	61.3	75.2	61.3

trast to the results shown in this figure, in systems lacking energy regeneration, nucleotide concentrations are depleted within just a few minutes. Although the predicted results exhibit a noticeable offset from the experimental data, the general trends and the order of magnitudes of the displayed concentration courses are in agreement with experiment. Furthermore, the model suggests an accelerated drop in ATP and GTP concentration, roughly within the initial 10 min of process time. Such a decrease is not mimicked by the corresponding experimental concentration curves. This observed discrepancy may be explained by a displacement of the binding equilibria for the system used at the start of the simulation, and are thus a result of the chosen initial conditions. In particular, the sum of the aminoacylation reactions (see Fig. 16d) appears to be responsible for the observed sharp decrease in NTP concentration. This finding may give some indication that the initial conditions for tRNA charging are probably over-estimated by the model.

Figure 17a plots the predicted rates for selected reactions of the energy regeneration network. The rates of both acetyl phosphate hydrolysis and ATPase reaction are found to decrease over time. On the other hand, the rates of acetate

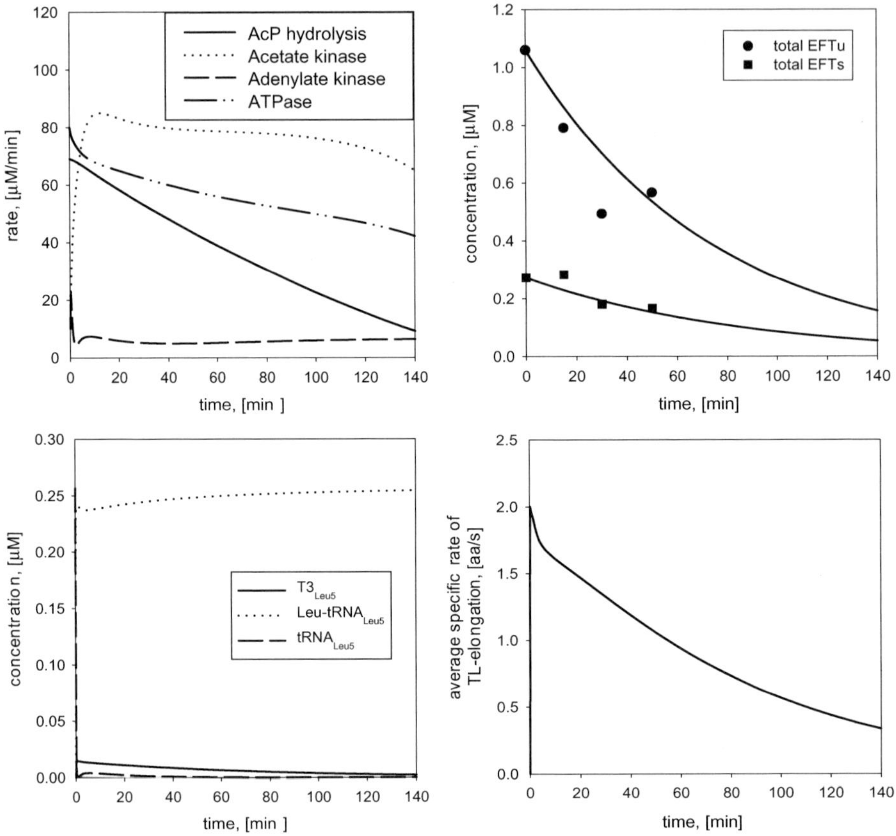

Fig. 17 Time courses of (**a**) predicted rates involved in energy consumption and regeneration, (**b**) measured and simulated total EFTu and EFTs levels (measurements were recomputed from Schindler et al. [138]), (**c**) predicted concentrations of $tRNA_{LeuS}$ in its uncomplexed form, aminoacylated state ($Leu\text{-}tRNA_{LeuS}$), and as ternary complex ($T3_{LeuS}$). Initial concentrations (at $t = 0$) were 0, 0, and 0.2566 μM for $T3_{LeuS}$, $Leu\text{-}tRNA_{LeuS}$, and $tRNA_{LeuS}$, respectively. (**d**) Predicted time course of average specific rate of translation elongation (per mRNA-bound ribosome). At $t = 0$, this rate is not defined (since there are initially no ribosomes bound to mRNA). It was ten taken to be equal to 0

kinase and adenylate kinase are shown to remain approximately constant over two hours of process duration. Hence, the endogenous energy regeneration system is shown to be capable of providing sufficient energy levels for at least two hours of process duration. This view is supported by the fact that the energy charge obtained from experimental data remained above 0.92 throughout the process (data not shown).

In Fig. 17b, the time-dependent trajectories of measured versus predicted total concentrations of the elongation factors EFTu and EFTs are illustrated. Both quantities show an exponential decay with time due to inactivation. The low absolute levels of these elongation factors are striking when compared to in vivo conditions. Under balanced growth, the concentrations of EFTu, EFTs, and EFG are (by factors of about 150, 20, and 20, respectively) higher than the initial conditions of the investigated in vitro system [101]. While the discrepancies for initial EFTs and EFG levels can be explained primarily by the dilution steps employed during lysate preparation, the preparation procedure apparently leads to a selective deprivation by EFTu concentration [138]. As production time progresses, the mismatch to ribosome concentration becomes increasingly severe, due to the noted inactivation of EFTu and EFTs, respectively.

The consequences of reduced EFTu levels are further reflected in Fig. 17c, where the simulated concentration courses of the various forms of $tRNA_{Leu5}$ are given versus time. The sum of the displayed concentrations together with the corresponding tRNA-species bound to elongating ribosomes add up to roughly 0.26 μM at any instant during the process time (there is no tRNA degradation considered here). As is obvious from this figure, the split ratio between Leu-$tRNA_{Leu5}$ and its corresponding ternary complex is very large. It increases from 16 to 115 over the course of the experiment. The predominant conformation in which this tRNA is predicted to exist is the aminoacylated form. This also holds true for the other 34 tRNA species considered (data not provided). In other words, this means a highly unfavorable situation for elongation kinetics, since tRNA is required as ternary complexes to serve as a substrate at each step of translation elongation. The average specific rate of ribosomal elongation, as sketched in Fig. 17d, is thus predicted to decline from about 2 aa/s to roughly 0.3 aa/s within almost 2.5 hours of experiment duration. On the other hand, in vivo, the average specific rate of peptide bond formation ranges between 10 to 20 aa/s [101]. Hence, an approximate 5 to 60-fold difference exists between specific protein synthesis rates obtained in vivo and the investigated in vitro system. These findings together strongly suggest the need for an appropriate supplementation of purified translation factors, most importantly of EFTu in this case, in order to maintain their catalytically active forms at levels necessary for efficient translation elongation.

The rates of mRNA synthesis and degradosome association are both depicted in Fig. 18a. With declining nucleotide concentrations and due to the modeled inactivation of the enzyme T7 RNA polymerase, the rate of transcription is found to diminish with time. However, it is shown to remain above the rate of degradosome association throughout the displayed time period. On the other hand, the rate of degradosome association increases with time. As can be viewed from the similarity to the time curve of mRNA concentration (see Fig. 16a), this rate is dictated by mRNA availability. The average specific rate of degradosome movement was predicted to be 31.7 codons/s in the investigated system and remained essentially constant across the entire process (data not shown).

After an initial experimental period of about 10 minutes, the predicted average gap between degradosomes settled at 690 codons (Fig. 18b). This means

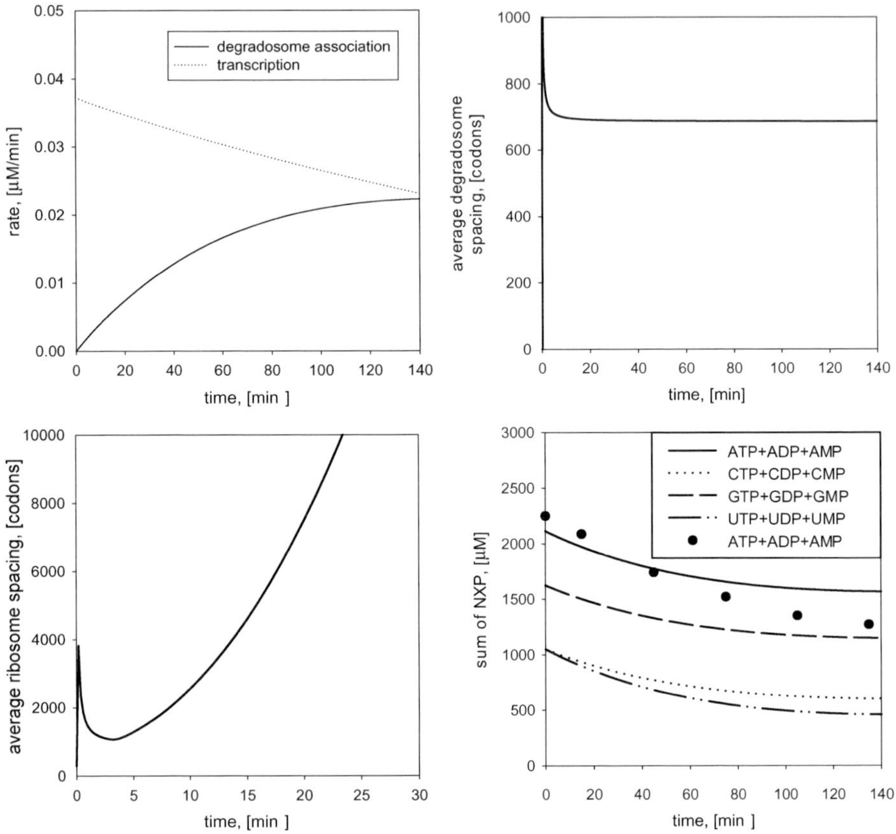

Fig. 18 Time courses of predicted (**a**) rates of transcription and degradosome asociation, (**b**) average spacing between mRNA-bound degradosomes, (**c**) spacing among mRNA-bound ribosomes, and (**d**) sum of concentrations of adenylates, cytidylates, guanylates, and uridylates, respectively. The measured total adenylate concentration is also given

that on average approximately one degradosome was bound per two molecules of full-length mRNA (consisting of 357 base triplets each). On the other hand, average ribosome densities indicated that, at the most, one ribosome was bound per three native mRNA transcripts. This situation corresponds to the local minimum of ribosome spacing at $t = 3$ min displayed in Fig. 18c. During subsequent process times, ribosome spacing was found to increase exponentially, in agreement with the exponential slow-down in translation initiation introduced into the model Eq. 103. The average distance of translating ribosomes was at all times during the process predicted to be greater than the average spacing between mRNA-bound degradosomes. At process termination after 140 min, there was only one ribosome bound per approximately 7000 mRNA molecules according to the model (data not shown). These values should be compared to average ribosome distances of about 40 to 80 codons in a growing *E. coli* cell [101], a factor of about 100 lower than predicted for the in vitro system.

In the above, the transcription rate was demonstrated to be able to compensate for the endogenous mRNA degradation processes. The choice of T7 RNA polymerase concentration added to the system even appears to be overdimensioned, since lower mRNA levels in conjunction with higher ribosome densities could have well been tolerated. Higher ribosome loadings can function as an effective protection mechanism against ribonucleolysis (Sect. 4). In fact, excessive mRNA levels may not be desirable, since mRNA synthesis is highly energy consuming. Further, the pool of transcripts constitutes a significant sink for nucleotides. Material balancing revealed that the reduction in total nucleotide levels matched the nucleotide requirements for generating the measured mRNA concentration (data not provided). Therefore, even in the presence of a functioning co-factor regeneration system, that pushes nucleotide concentrations to their most phosphorylated state, the total sum of nucleotides is also noted to decrease with time (see Fig. 18d). Hence, the noted drop in the concentrations of both ATP and GTP (see Fig. 16c), as well as CTP and UTP (data not shown), can be explained with their incorporation into mRNA, instead of them being degraded.

Low ribosome densities imply negligible sterical effects among translating ribosomes. This is in agreement with ribosomal queueing factors being predicted to be close to unity. As a representative constituent of all queueing factors for translation elongation, the time course of factor q_{14}^{R} is displayed in Fig. 19a. This factor remains almost equal to 1 throughout the process. The only exception among all queueing factors where a significant difference from 1 was observed, at least temporarily in this study, is the queueing factor for translation initiation (q_{22}^{R0}, depicted in Fig. 19a). This factor, denoting the probability of the ribosome binding site being unoccupied, is shown to increase from about 0.80 at simulation start to a value of about 1 within the initial 10 minutes of process time. During this time interval, the concentration of mRNA is low, so that the fraction of occupied ribosome binding sites is greater than at subsequent process times, which corresponds to higher mRNA levels.

When investigating the dynamics involved in the loading process of an initially naked mRNA, interesting phenomena can be noted. As is visualized in Figure 19b, the rates of translation initiation, elongation, and termination are shown to increase initially, as ribosomes are loaded onto the (previously naked) mRNA. Elongation rates at codons 107 and 207 (as well as at the termination site (codon 273)) show a time-delayed response, which corresponds to the time gap needed for ribosomes to travel the distance between the initiation site and the respective codon (codons 107, 207, and 273). The trajectories of the rates of 70S initiation complex formation and IF2-dissociation are indistinguishable in this graph. Both of these rates reach a maximum when the contribution from the inactivation of ribosomal protein S1 just equals the effect of substrate availability on 70S initiation complex formation rate, and are found to drop afterwards.

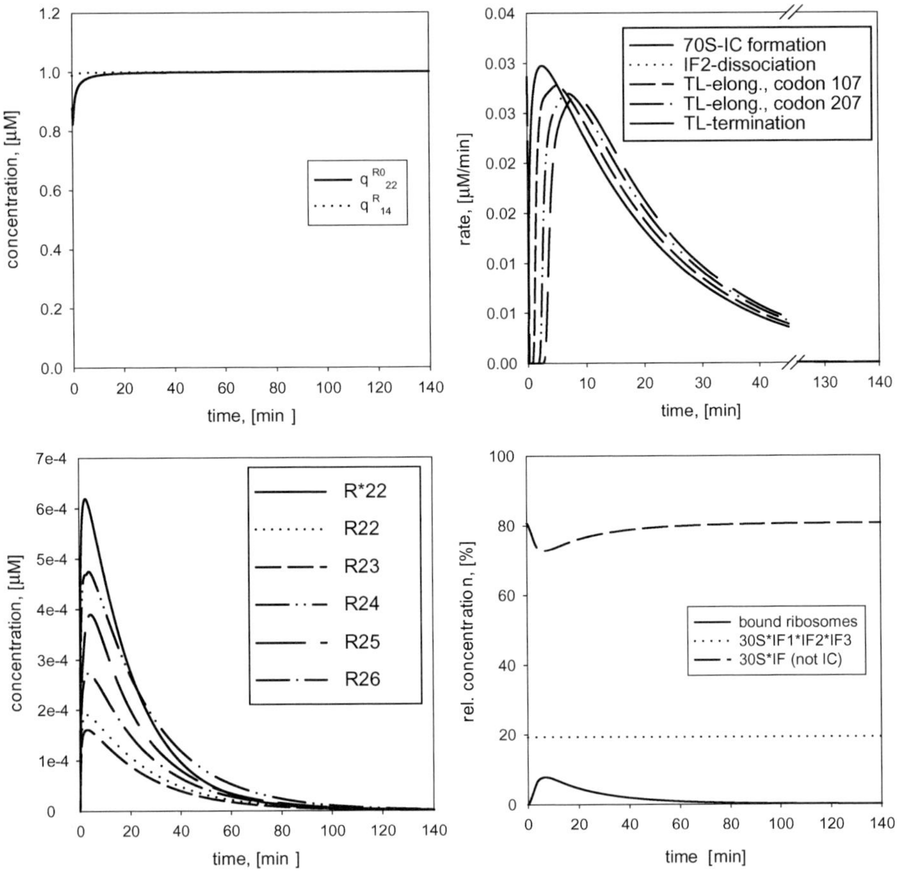

Fig. 19 (**a**) Predicted time courses for two selected queueing factors. q_{22}^{R0} denotes the probability of the ribosome binding site being unoccupied. q_{14}^{R} represents the probability of forward movement onto codon 15 (**b**) Predicted time courses for rates of translation initiation, elongation, and termination (**c**) Simulated time courses for concentrations of mRNA-bound ribosomes at selected codons in the vicinity of the start codon (number 22). Symbols R*22 and R22 distinguish ribosomes bound to the initiation codon prior and subsequent to IF2-dissociation, respectively (**d**) Predicted time courses of relative ribosome concentrations

The step-wise propagation of ribosomes along the mRNA causes temporally-spaced processes to take place, which are, for example, reflected in the codon-specific elongation rates. Viewing the trajectory of each elongation rate as a frequency distribution, the mean of the distribution moves to higher values with increasing codon number, while the profile is smoothed. This is a behavior generally observed for Poisson distributions, as was pointed out earlier [34, 35].

Figure 19c shows the concentrations of ribosomes bound to the initiation codon (number 22), and to codon positions immediately after the start codon.

As can be seen from this figure, the concentration of ribosomes representing 70S-initiation complexes (symbol R*22) is shown to be higher than the concentration of ribosomes that are bound to this position after IF2-dependent GTP hydrolysis (state R22). Ribosomes occupying the initiation site thus effectively function as a road-block, in the sense that they prevent upstream propagating degradosomes from getting access to endonucleolytic cleavage sites contained within the coding region.

Furthermore, it should be mentioned that time profiles displayed in Fig. 19c are not exactly Poisson-distributed. This follows as a direct consequence of variable codon-specific elongation rates. Ribosomal loading patterns will thus evolve that compensate for these codon-specific differences. Explicitly this means that codons corresponding to relatively lower specific elongation rates will show higher ribosome loadings, in order to maintain volumetric elongation rates that are equal for all codons j during pseudo-steady state synthesis conditions.

In Fig. 19d, the predicted relative concentrations of ribosomes bound to mRNA, ribosomal subunits bound to all three initiation factors simultaneously (symbol 30S·IF1·IF2·IF3), and the remainder of 30S subunits (freely dissolved and complexed with any one or multiple, but not all initiation factors simultaneously) are plotted. The sum of these three quantities adds up to 1 at any process time, since total ribosome concentration is considered invariant here. Over the entire time course, about 80% of all ribosomes are predicted to be in a state that is neither bound to mRNA, nor complexed at the same time with all three initiation factors. The time profile of this pool shows a slight drop within roughly the initial 20 minutes, as ribosomes get loaded onto mRNA. Most noticeably, the concentration of complex 30S·IF1·IF2·IF3 stays virtually unaffected by the dynamics of translation. It takes a value of about 20% of the total ribosome concentration. The concentration of 30S·IF1·IF2·IF3, however, influences the rate of 70S initiation complex formation in a linear fashion (Sect. 5). The equilibrium between 30S·IF1·IF2·IF3 and the non-active forms of 30S (complexed with less than all three initiation factors) could be favorably shifted at higher levels of initiation factors, so that ideally all ribosomes unbound to mRNA would exist as complex 30S·IF1·IF2·IF3. In this case, the initial volumetric rate of protein synthesis could theoretically be raised by a factor of 5 at the most, unless further rate limitations exist.

6.5 Optimization of Translation Factor Levels

One of the results obtained from simulating cell-free GFP production in the previous section was that dilute translation factor levels were predicted to be the primary cause of the low protein production rates observed. In order to further investigate this hypothesis and to check whether higher total translation factor levels would lead to a performance improvement, the previously described model was subjected to a sequence of raised initial concentrations of total translation factors, and the resulting system dynamics were simulated. The reference to which elevated initial translation factor concentrations are compared, is the same as for the cell-free protein synthesis system described in Sect. 6.4.

In the following analysis, the impact of selectively increasing (A) the concentration of elongation factor EFTu, (B) the concentrations of all elongation factors simultaneously, (C) the concentration of all initiation factors, and (D) the concentrations of all initiation factors and elongation factors considered at the same time was investigated. The initial conditions of the respective simulations are compared in Table 8. Importantly, all other reaction conditions and initial concentrations were kept the same as in the reference system. The time-dependent inactivation of selected compounds identified earlier was also considered here.

6.5.1 Effect of Elongation Factor Concentration

Figure 20 shows predicted time traces for the average specific rate of translation elongation for various total EFTu concentrations. As can be seen from this graph, increasing the level of EFTu is predicted to lead to a significant enhancement in average specific ribosome propagation rate. Doubling the EFTu concentration at the start of simulation is predicted to give a higher (by a factor of 1.8) average specific elongation rate at $t = 0$ (dotted line) than for the reference condition (solid line). This finding indicates an almost 1 : 1 improvement and suggests that in the earlier scenario, EFTu concentration was indeed limiting this rate. At EFTu levels equal to and higher than (by a factor of 20) the reference system (Sect. 6.4), the average rate of ribosome elongation is predicted to reach a maximum of 11.5 aa/s. This rate lies within the range of in vivo specific rates of peptide bond formation (10 to 20 aa/s). Thus, by increasing EFTu concentration, the stringent limitations on specific elongation rate noted earlier could in theory be successfully overcome, until further rate-limitations begin to apply (that set the upper-boundary threshold shown in Fig. 20).

When the initial levels of elongation factors EFG and EFTs were raised by a factor of 30 in addition to EFTu concentration (scenario B in Table 8), no further performance improvement was noted. The final concentration of protein product, as well as translation initiation rate, the specific rate of translation elongation, and the fractional splitting among ribosomes were all predicted to be the same as for the system with increased EFTu concentration only (see Table 9).

Notably, time profiles for the concentration of protein product GFP are the same for systems with raised EFTu concentrations only and for the system where all EF concentrations were raised simultaneously (data not pro-

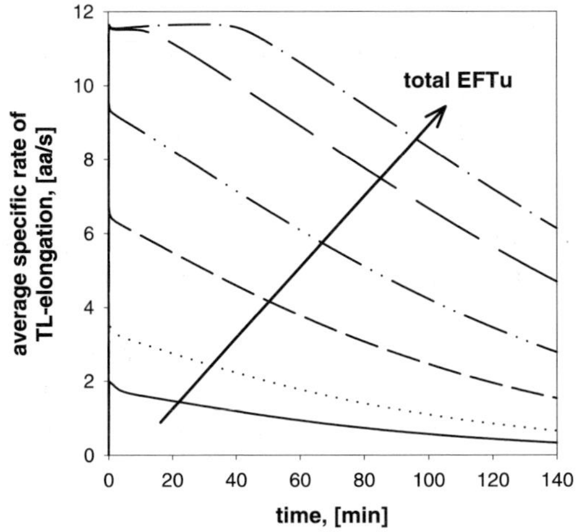

Fig. 20 Impact of EFTu concentration on the average specific rate of translation elongation (per mRNA-bound ribosome). The *solid line* is replotted from Fig. 17d. The other trajectories correspond to the initial total EFTu concentration increased by factors of 2, 5, 10, 20, and 30, respectively, in comparison to the reference conditions described in Sect. 4

Table 9 Results from simulating cell-free protein synthesis during the optimization of translation factor concentrations. C_{Prot} is the protein concentration at $t = 140$ min. Other quantities displayed were taken at time $t = 2$ min, respectively. All of these quantities remained essentially constant throughout the process, except for the average specific rate of elongation $(k_{\text{TLE}})_{\text{avg}}$, which decreased with the process time. A – 30-fold EFTu concentration in comparison to the reference state. B – All EF concentrations are raised by a factor of 30, respectively. C – Raised IF levels. D – Simultaneous increase in the concentrations of both initiation factors and elongation factors

Condition	C_{Prot} (μM)	$V_{TLI,70\text{SIC}}$ (μM/min)	$(k_{\text{TLE}})_{\text{avg}}$ (aa/s)	$\frac{C_{30\text{S}\cdot\text{IF1}\cdot\text{IF2}\cdot\text{IF3}}}{C^{\text{R}}_{\text{tot}}}$ (%)	$\frac{C^{\text{R}}_{\text{bound}}}{C_{70\text{S}_{\text{tot}}}}$ (%)
Reference	0.69	0.03	1.9	19.3	3.4
A: EFTu	0.70	0.03	11.5	19.3	0.8
B: EF	0.70	0.03	11.5	19.3	0.8
C: IF	n.d.	0.10	1.7	85.2	11.2
D: IF + EF	3.17	0.13	10.7	88.5	4.1

vided). They are all virtually identical to the time profile of synthesized GFP that is displayed in Fig. 16a. Also, the final concentration of protein product achieved after 140 minutes of process time is predicted to be virtually identical (equal to 0.70 μM) across all the different systems with elevated EF concentrations. The effect of raising total EF concentration was exclusively an increased specific translation elongation rate. This finding simply means that elongating ribosomes travel faster along the mRNA under conditions of raised EF concentration. The number of mRNA-bound ribosomes remains, however, unchanged from the system of non-elevated EF concentration, and the same number of GFP molecules is completed per unit of time.

As demonstrated, an enhancement of *specific* protein synthesis rate is not necessarily sufficient to also ensure improved *volumetric* protein production rates. Raising volumetric productivity is generally achieved by increasing catalyst levels. In the case of protein synthesis, this is equivalent to driving ribosomes to a mRNA-bound state. Higher ribosome densities are expected to occur at higher rates of translation initiation. Due to the previously-noted excess of freely dissolved ribosomes in this study in contrast to their active form as a complex with initiation factors, raised IF concentrations are expected to yield higher rates of translation initiation. Thus, the impact of increasing the initiation factor concentration on protein synthesis rate is examined in next section.

6.5.2 Effect of Initiation Factor Concentration

An improvement in volumetric protein production rate was suggested to be obtained by raising initiation factor levels in an appropriate stoichiometric ratio to total ribosome concentration. This working hypothesis was subsequently tested by simulating cell-free protein synthesis dynamics with raised initial concentrations of initiation factors (condition C in Table 8).

Under these conditions, an improved rate of 70S initiation complex formation is indeed noted. This translation initiation rate of 0.10 μM/min is predicted to be 3.5-fold higher than the corresponding rate of the reference simulation (0.03 μM/min) (see also Table 9). As can be viewed from further data provided in this table, the enhancement can be explained by a favorable shift of non-translating ribosomes towards full complexation with all three initiation factors considered (an increase from 19.3% to 85.2%). This compound influences the rate of 70S initiation complex formation linearly (Sect. 5). Interestingly, however, numerical integration was only found to cease after 4 min of simulated process time. In the situation applied here, the ribosomes showed a tendency to stall when bound to mRNA, due to a lack of sufficient amounts of elongation factors that would promote the rate of translation elongation. Apparently, sterical interactions among translating ribosomes were found to propagate backwards to the ribosome binding site (data not provided), which ultimately led to premature simulation termination. This finding indicates that at higher rates of translation initiation, sufficiently high specific rates of translation elongation become increasingly important, because they can ensure a sufficiently high rate of clearance of the ribosome binding site.

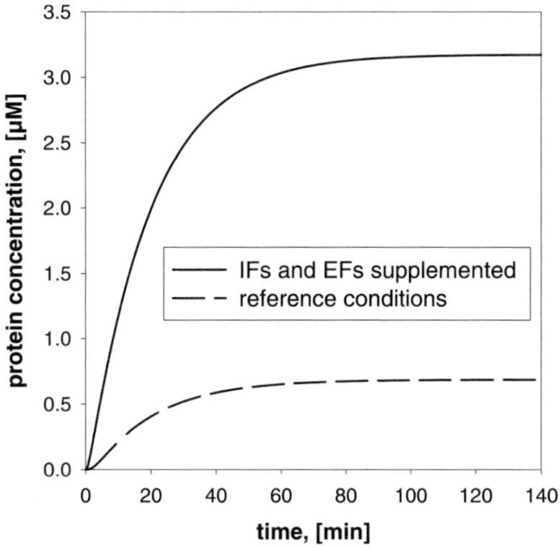

Fig. 21 Time profile of protein concentration under reference conditions and for a system with combined supplementation of initiation factors (IF1, IF2, and IF3) and elongation factors (EFTu, EFG, and EFTs)

Consequently, in the next step of the opimization strategy, the concentrations of initiation factors and elongation factors were raised simultaneously (scenario D in Table 8). Under these conditions, a tremendous improvement in cell-free protein synthesis was predicted. Figure 21 shows a comparison between the predicted product protein concentration vs time profile for the reference simulation with the profile observed for the situation where the levels of translation initiation and elongation factors were optimized. As can be seen from this figure, the final concentration of protein product was predictted to reach a level of 3.17 µM (in contrast to 0.69 µM obtained in the reference system). The initial rate of translation initiation was 0.13 µM/min (compared to the reference rate of 0.03 µM/min). The concentration of 30S ribosomal subunits that exists in a complex with all initiation factors taken into account simultaneously is calculated to be 88.5% in this case (19.3% in the reference system). All three quantities, C_{Prot}, $V_{\mathrm{TLI,70SIC}}$, and the fractional amount of complex 30S·IF1·IF2·IF3, showed a 4.6-fold increase in comparison to the reference condition (Table 9). In this case, the average specific rate of translation elongation was predicted to be 10.7 aa/s, which falls within in vivo levels (10 to 20 aa/s).

In summary, the model predicts that only a combination of simultaneously increasing the levels of both translation initiation and elongation factors significantly improves both specific and volumetric protein production rates in comparison to the chosen reference state.

7 Conclusions

In this study, a dynamic model of prokaryotic gene expression was developed that makes substantial use of gene sequence information. The main contribution arises from the fact that the combined gene expression model allows us to assess the impact of nucleotide sequence alteration on the dynamics of gene expression rates mechanistically. The high level of detail of the mathematical model enables us to provide a highly detailed insight into the various steps of the protein expression process.

Modeling required the development of a valid model structure for template-bound biopolymerization processes within a continuous analysis method. In contrast to a discrete model, or a combination of both approaches (hybrid modeling), the continuous model presented is a mechanism-based deterministic description of system states in terms of differential and algebraic sets of equations. Characteristically, a codon-specific representation of state variables was chosen for this model.

Transcription kinetics were described mathematically for the example of T7 RNA polymerase. Parametrization of the transcription model was carried out for selected model constants (for the rate constants of initiation, elongation, and termination), as well as for the maximum rate of transcription typical reaction.

The process of mRNA degradation was modeled allowing for a distinction between endonucleolytic and exonucleolytic reaction steps. The effects of increased translational efficiency, greatly improving mRNA stability, as observed experimentally, were correctly demonstrated by the model. By simulating *lacZ* mRNA degradation, it was possible to identify the parameters contained in the degradation model.

Because mRNA can constitute a significant sink for nucleoside triphosphates, it was proposed that the transcription rate should be kept at moderate levels, in particular in batch systems. Otherwise, the resulting nucleotide concentrations may drop to limiting thresholds as they are incorporated into mRNA molecules. Model-assisted simulations can help to identify an appropriate counterbalance between mRNA degradation rate and a suitable transcription rate.

The translation model presented covers the mechanisms of protein synthesis initiation, elongation, and termination, at the same time considering the particular mechanistic roles of key translation factors. An earlier approach to describing steric interference among template-bound catalyst [34] was extended in this study, in order to also cover a situation where two different types of catalysts (ribosomes and degradosomes) can be bound in multiple copies to the same template.

To enhance the applicability of the model to large expression systems, a reduced model was introduced. In the suggested procedure, the number of state variables were significantly diminished by merging groups of base triplets together, while at the same time taking into account the implications of this on reaction kinetics and material balancing.

The current status of the combined model allowed us to reveal several causes of production limitation: substrate depletion or inactivation processes, or unfavourable initial catalyst concentrations and their stoichiometric relations. An application of the combined gene expression model to simulating cell-free protein synthesis dynamics demonstrated that limited volumetric productivitites are caused by unfavourably low translation factor levels that are typical of these dilute in vitro systems. Equilibrium binding calculations suggested a requirement for at least equal molar ratios of initiation factors IF1, IF2, and IF3, with respect to the total concentration of unbound ribosomes. When these conditions are met, about 85% of all freely dissolved 30S ribosomal subunits are predicted to prevail in their activated form, in other words they are complexed with all of these three initiation factors. By raising the concentrations of both translation initiation and elongation factors appropriately, a four-fold improvement in volumetric protein synthesis rate and a five-fold higher final product yield are predicted over a non-optimized reference batch process.

From the standpoint of reduced model complexity, it may be beneficial to use the overall model to estimate mechanism-related parameters or decay constants of a gene expression model, prior to applying these parameters within a whole system modeling framework. The immediate value of such models arises from their ability to describe the expression of individual genes or a few genes at a time, which is typical for recombinant protein production.

Gene sequence information enters the overall model at the following stages: (a) within the transcription process, by assigning different rate constants for initiation and termination of mRNA synthesis, respectively, (b) the endo- and exonuclease activities in the ordered process of 5′ to 3′-degradation of messenger RNA, (c) during translation, by distinguishing codon-specific elongation rates and effects related to steric interactions among translating ribosomes.

In summary, the mathematical gene expression model presented in this study provides a comprehensive framework for a thorough analysis of sequence-related effects during mRNA synthesis, mRNA degradation, and ribosomal translation, as well as their nonlinear interconnectedness, and may therefore prove useful in the rational design of recombinant bacterial protein synthesis systems.

Acknowledgements Financial support by the German Ministry of Research (ZSP project A3.10U) and by the German Research Foundation (DFG project RE 632/8-1) is gratefully acknowledged. This project was also supported by the Federal Ministry of Education (BMBF) associated with joint project "Cell-free protein biosynthesis reactor" (project FKZ 0 311 302). We thank Volker Erdmann (Institute of Biochemistry, FU Berlin, Germany),

Alexander Spirin (Institute for Protein Research, Pushchino, Russia), Herbert Stadler (Insitute for Bioanalytics, Göttingen, Germany) and our industrial collaboration partner Roche Diagnostics Ltd. (Penzberg, Germany), represented by Albert Röder, for stimulating discussions.

Appendix

A
Derivation of Queueing Factors for Systems with Two Catalysts

The following paragraphs provide an extension of a model previously suggested by the working group of Gibbs for template-directed and enzyme-catalyzed polymerization [33–35]. In the original study, sterical interactions among template-bound catalysts of the same type were considered. In this study, an analogous derivation of these probabilities is given for the case of two types of catalysts (in multiple copies) bound to the same template. Further new aspects of this model arise due to the transition from a fractional system description to one employing molarities, and due to the resulting consequences for material balancing.

A.1
Nomenclature

Parameters m_{D} (with $1 \le m_{\mathrm{D}} \le L_{\mathrm{D}}$) and m_{R} (with $1 \le m_{\mathrm{R}} \le L_{\mathrm{R}}$) characterize the positions of the catalytic center for catalysts D and R, respectively (see Fig. 5). If a site j is covered by catalyst D, its surrounding $j - m_{\mathrm{D}} + 1, \ldots, j - m_{\mathrm{D}} + L_{\mathrm{D}}$ sites are simultaneously blocked by this catalyst. Similarly, catalyst R covers L_{R} sites at a time within the vicinity of its binding site. Overlapping of catalysts is excluded.

The relative positions of a catalyst, while site j is in different states, are explained in Fig. 22. A site j on the template can be either empty (state $s = 0$), or in L_{D} different states of catalyst D, or L_{R} different states of catalyst R. In total, that makes $L_{\mathrm{D}} + L_{\mathrm{R}} + 1$ different states s for each site. The fractional occupancy of site j occupied by catalyst D that is in state s is given by $n_j^{(s)}$. The fractional occupancy of this site with respect to catalyst R in state s is denoted by $\tilde{n}_j^{(s)}$. The summation over all the states for site j leads to unity, according to

$$n_j^{(0)} + \sum_{s=1}^{L_{\mathrm{D}}} n_j^{(s)} + \sum_{s=1}^{L_{\mathrm{R}}} \tilde{n}_j^{(s)} = 1\,. \tag{106}$$

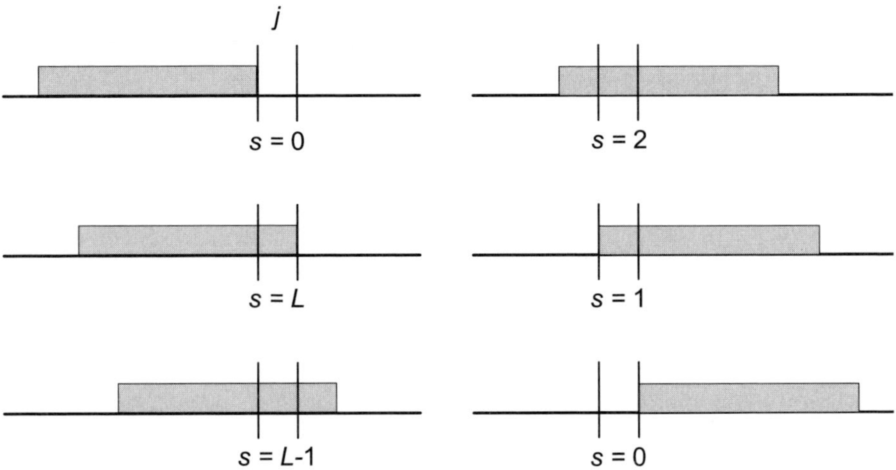

Fig. 22 Defining the different states a template-bound catalyst can take

A.2
Probabilities for Unoccupied Sites

Site $j+1$ can be empty only if site j is either in state 0, L_D, or state L_R, but not otherwise. Any other state s would cause a blocking of position $j+1$ and thus preclude catalyst movement onto this site. If site j is in either of the states 0, L_D, or L_R, site $j+1$ must take one of exactly three states: site $j+1$ is in this case either unoccupied ($s=0$), or in state 1 of either of the two catalysts.

Individual states of site j are distinguished together with the restrictions consequently imposed on site $j+1$. If site j is in state 0, then there are at the same time only three states possible for site j+1, namely in this case either empty ($s=0$), or state 1 of catalyst D, or else state 1 of catalyst R. It follows that if site j is in state L_D or L_R, then site $j+1$ can only take any one of the three states, either 0 or 1 for either of the two catalysts. Thus, if site j is in any one of the states, 0, L_D, or L_R, respectively, then at the same time, site $j+1$ needs to be in any one of the three states 0 or 1 for catalysts D and R, respectively. The converse is true, too. This leads to the following relation:

$$n_j^{(0)} + n_j^{(L_D)} + \tilde{n}_j^{(L_R)} = n_{j+1}^{(0)} + n_{j+1}^{(1)} + \tilde{n}_{j+1}^{(1)} \,. \tag{107}$$

The sum of fractional loadings of site j in states 0, L_D, and L_R just equals the sum of fractions in states 0 and 1 of site $j+1$. Under the assumption that no causal relationship exists for site $j+1$ to be empty whether site j is in state L_D, or L_R, or empty itself [35], the conditional probability, q_j, that site $j+1$ is empty may be expressed as

$$q_j = \frac{n_{j+1}^{(0)}}{n_{j+1}^{(0)} + n_{j+1}^{(1)} + \tilde{n}_{j+1}^{(1)}} . \tag{108}$$

Considering Eq. 106, Eq. 108 yields

$$q_j = \frac{1 - \sum_{s=1}^{L_D} n_{j+1}^{(s)} - \sum_{s=1}^{L_R} \tilde{n}_{j+1}^{(s)}}{1 - \sum_{s=1}^{L_D} n_{j+1}^{(s)} - \sum_{s=1}^{L_R} + \tilde{n}_{j+1}^{(s)} + n_{j+1}^{(1)} + \tilde{n}_{j+1}^{(1)}} . \tag{109}$$

A transformation of variables leads to an expression for the state s relative to the states L_D and L_R, respectively:

$$n_j^{(s)} = n_{j-s+L_D}^{(L_D)} \quad \text{for} \quad 1 \leq s \leq L_D \tag{110}$$

$$\tilde{n}_j^{(s)} = \tilde{n}_{j-s+L_R}^{(L_R)} \quad \text{for} \quad 1 \leq s \leq L_R . \tag{111}$$

With Eqs. 110 and 111, it can be shown that the following relation holds for $1 \leq s \leq L_D$, and $1 \leq s \leq L_R$, respectively:

$$\sum_{s=1}^{L_D} n_j^{(s)} = \sum_{s=1}^{L_D} n_{j-s+L_D} = \sum_{s=1}^{(L_D)} n_{j+s-1}^{(L_D)} \tag{112}$$

$$\sum_{s=1}^{L_R} \tilde{n}_j^{(s)} = \sum_{s=1}^{L_R} \tilde{n}_{j-s+L_R} = \sum_{s=1}^{(L_R)} \tilde{n}_{j+s-1}^{(L_R)} . \tag{113}$$

Equation 109 can then be rewritten in terms of the states L_D and L_R:

$$q_j = \frac{1 - \sum_{s=1}^{L_D} n_{j+s}^{(L_D)} - \sum_{s=1}^{L_R} \tilde{n}_{j+s}^{(L_R)}}{1 - \sum_{s=1}^{L_D-1} n_{j+s}^{(L_D)} - \sum_{s=1}^{L_R-1} \tilde{n}_{j+s}^{(L_R)}} . \tag{114}$$

For arbitrary reference states, m_D (with $1 \leq m_D \leq L_D$) and m_R (with $1 \leq m_R \leq L_R$), Eq. 114 reads

$$q_j = \frac{1 - \sum_{s=1}^{L_D} n_{j+s}^{(m_D)} - \sum_{s=1}^{L_R} \tilde{n}_{j+s}^{(m_R)}}{1 - \sum_{s=1}^{L_D-1} n_{j+s}^{(m_D)} - \sum_{s=1}^{L_R-1} \tilde{n}_{j+s}^{(m_R)}} . \tag{115}$$

Strictly speaking, Eq. 114 is only valid for the particular situation that $L_D = L_R$ and $m_D = m_R$. In this case, q_j is the same for either of the two catalysts. On the other hand, if both catalysts show a divergence in lengths (when $L_D \neq L_R$), and when they have different reference states ($m_D \neq m_R$), q_j will differ with respect to the type of catalyst. This is demonstrated later. First, q_j^D, is derived for catalyst D, before this term is elaborated analogously for catalyst R.

For convenience, L_D and L_R are assumed to fulfill the condition that $L_D < L_R$. It may be further imposed that $m_D = m_R = 1$. These assumptions can be abandoned later on. A movement of catalyst D located in site j to position $j+1$ is impeded by of all the catalysts that are bound (with respect to their reference state) throughout the sites $j+1$ to $j+L_D$. All other catalysts whose reference states are located beyond this interval (at sites greater than $j+L_D$, or at sites smaller than j) do not affect the movement of D from site j into site $j+1$. In particular, this means that the catalysts R bound to sites L_D+1 to L_R, obviously cause no impact on the queueing of catalyst D. This may be taken into account when mathematically describing q_j for catalyst D. If additionally the assumption of equal reference states is dropped, so that $m_D \neq m_R$ is permitted, Eq. 115 may thus be modified to yield for catalyst D

$$q_j^D = \frac{1 - \sum_{s=1}^{L_D} n_{j+s}^{(m_D)} - \sum_{s=1}^{L_D} \tilde{n}_{j+s}^{(m_R)}}{1 - \sum_{s=1}^{L_D-1} n_{j+s}^{(m_D)} - \sum_{s=1}^{L_D-1} \tilde{n}_{j+s}^{(m_R)}} . \tag{116}$$

From now, the superscript indicating the reference state is neglected. Queueing factors for catalysts D and R located in position j, respectively, can be rewritten in the following form:

$$q_j^D = \frac{1 - \sum_{s=1}^{L_D} n_{j+s} - \sum_{s=1}^{L_D} \tilde{n}_{j+s-m_D+m_R}}{1 - \sum_{s=1}^{L_D-1} n_{j+s} - \sum_{s=1}^{L_D-1} \tilde{n}_{j+s-m_D+m_R}} \tag{117}$$

$$q_j^D = \frac{1 - \sum_{s=1}^{L_R} n_{j+s-m_R+m_D} - \sum_{s=1}^{L_R} \tilde{n}_{j+s}}{1 - \sum_{s=1}^{L_R-1} n_{j+s-m_R+m_D} - \sum_{s=1}^{L_R-1} \tilde{n}_{j+s}} . \tag{118}$$

Equations 117 and 118 denote the probabilities that site $j+1$ is accessible when the respective catalyst (D or R) is bound to site j.

A.3
Catalyst Association

Similarly, the previously-derived probability for catalyst association (MacDonald and Gibbs [35]) needs to be modified in order to accomodate a situation where two different types of catalysts are considered. In this case, the binding site (j_{D0}) for catalyst D may not coincide with the binding location for R (j_{R0}). For example, it may be assumed that $j_{D0} < j_{R0}$. That is, catalyst D is taken to bind further upstream than R. In this case, the binding of catalyst R would be hampered not only by the catalysts bound to sites j with $j_{R0} \leq j \leq j_{R0} + L_R$, but also by catalyst D bound within $L_D - 1$ sites upstream from j_{R0}. If this additional interaction is taken into consideration, and without fixing the positional order of binding a priori, the probabilities for unoccupied binding sites can thus be derived for catalysts D and R, respectively. That is,

$$q_j^{D0} = 1 - \sum_{s=1}^{L_D} n_{j_{D0}+s-1} - \sum_{s=1}^{L_D+L_R-1} \tilde{n}_{j_{D0}+s-m_D-L_R+m_R} \tag{119}$$

$$q_j^{D0} = 1 - \sum_{s=1}^{L_D+L_R-1} n_{j_{R0}+s-m_R-L_D+m_D} - \sum_{s=1}^{L_R} \tilde{n}_{j_{R0}+s-1} \,. \tag{120}$$

A.4 Transition to Concentrations

When the fractional notation is substituted by the concentrations of state variables involved in mRNA degradation, the following set of equations can be obtained. For degradosome association, which occurs at base triplet $j_{D0} = m_D$, the probability of this site being unblocked depends on the concentrations of both the degradosomes and the ribosomes bound to the vicinity of this site. $q_{j_{D0}}^{D0}$ is thus expressed by

$$q_{j_{D0}}^{D0} = 1 - \sum_{s=1}^{L_D} \frac{\sum_i C_{i,jD0+s-1}^{D}}{C_{jD0+s-1}{}^{M}} - \sum_{s=1}^{L_D+L_R-1} \frac{C_{jD0+s-m_D-L_R+m_R}}{C_{jD0+s-m_D-L_R+m_{R0}}^{M}} \,. \tag{121}$$

Degradosome movement along a mRNA is influenced by both degradosomes and ribosomes bound to nearby sites downstream of a base triplet j. The probability of site $j+1$ being empty is given by

$$q_j^{D} = \frac{1 - \sum_{s=1}^{L_D} \frac{\sum_i C_{i,j+s}^{D}}{C_{j+s}^{M}} - \sum_{s=1}^{L_D} \frac{C_{j+s-m_D+m_R}^{R}}{C_{j+s-m_D+m_R}^{M}}}{1 - \sum_{s=1}^{L_D-1} \frac{\sum_i C_{i,j+s}^{D}}{C_{j+s}^{M}} - \sum_{s=1}^{L_D-1} \frac{C_{j+s-m_D+m_R}^{R}}{C_{j+s-m_D+m_R}^{M}}} \tag{122}$$

with $j_{D0} \leq j \leq J$. Analogously, the queueing factor for ribosome association at the initiation codon $j = j_{R0}$ is affected by both ribosomes and degradosomes covering this site. That is,

$$q_{j_{R0}}^{R0} = 1 - \sum_{s=1}^{L_D+L_R-1} \frac{\sum_i C_{i,j_{R0}+s-m_R-L_D+m_D}^{D}}{C_{j_{R0}+s-m_R-L_D+m_D}^{M}} - \sum_{s=1}^{L_R} \frac{C_{j_{R0}+s-1}^{R}}{C_{j_{R0}+s-1}^{M}} \,. \tag{123}$$

The queueing factor for translational elongation, q_j^{R} (with $j_{R0} \leq j \leq K$), describes a dependency on both the neighboring degradosome concentration and that of the ribosomes, according to

$$q_j^{R} = \frac{1 - \sum_{s=1}^{L_R} \frac{\sum_i C_{i,j+s-m_R+m_D}^{D}}{C_{j+s-m_R+m_D}^{M}} - \sum_{s=1}^{L_R} \frac{C_{j+s}^{R}}{C_{j+s}^{M}}}{1 - \sum_{s=1}^{L_R-1} \frac{\sum_i C_{i,j+s-m_R+m_D}^{D}}{C_{j+s-m_R+m_D}^{M}} - \sum_{s=1}^{L_R-1} \frac{C_{j+s}^{R}}{C_{j+s}^{M}}} . \tag{124}$$

The summation over index i used in Eqs. 121 and 124 denotes the sum of degradosomes in different conformations bound to a codon j, according to

$$\sum_i C_{i,j}^{D} = C_j^{D^*} + C_j^{D^*\mathrm{Frag}} + C_j^{D} . \tag{125}$$

Given the finite dimensions of a degradosome, degradosome binding to base triplets upstream of j_{D0} is excluded, thus

$$C_j^{D} = 0 \quad \text{for} \quad j < j_{D0} . \tag{126}$$

Further, ribosome binding within non-coding regions is neglected. This yields,

$$C_j^{R} = 0 \quad \text{for} \quad j < j_{R0} \quad \text{and} \quad K < j \leq J . \tag{127}$$

B
Derivation of Enzymatic Rate Equations

Kinetic rate expressions were derived with the method and program described in [147]. Rate derivation is based exclusively on the pseudo-steady state condition and the assumption of rapid equilibrium.

B.1
70S Initiation Complex Formation

Using symbols $[E]_t$ = total concentration of complex 30S·IF1·IF2·GTP·IF3, $[A]$ = concentration of fMet-tRNA$_f^{M}$, $[B]$ = concentration of ribosome binding sites, $[C]$ = concentration of ribosomal subunit 50S, $[P]$ = concentration of IF1,

$[Q]$ = concentration of IF3, the elementary reaction steps read:

$$E + A \overset{K_A}{\rightleftharpoons} EA \tag{128}$$

$$EA + B \overset{K_B}{\rightleftharpoons} EAB \tag{129}$$

$$E + B \overset{K_N}{\rightleftharpoons} EB \tag{130}$$

$$EB + A \overset{K_A}{\rightleftharpoons} EAB \tag{131}$$

$$EAB \overset{K_{TLI,70SIC,1}}{\rightleftharpoons} EPQ \tag{132}$$

$$EPQ + C \overset{K_{TLI,70SIC,2}}{\rightleftharpoons} E + P + Q\,. \tag{133}$$

From Eqs. 128 to 133, the following rate equation was derived:

$$V_{TLI,70SIC} = \frac{k_{TLI,70SIC,1}\, k_{TLI,70SIC,2}[A][B][C][E]_t}{D} \tag{134}$$

with

$$D = k_{TLI,70SIC,2}[C](K_A\, K_B + K_B[A] + K_A[B] + [A][B]) + k_{TLI,70SIC,1}[A][B]\,.$$

B.2 Translation Elongation

Symbols are $[E]_t$ = total concentration of ribosomes bound to mRNA at codon j; $[A]$, $[C]$, $[D]$ = concentrations of ternary complexes ($T3_j$); $[B]$ = concentration of EFG·GTP; $[P]$ = concentration of P_i; $[Q]$ = concentration of EFTu·GDP; $[R]$, $[M]$, $[O]$ = concentrations of tRNA species, and $[T]$ = concentration of EFG·GDP. The elementary reaction steps spanning $n_c = 3$ consecutive elongation cycles are represented by:

$$E + A \underset{k_{-1}}{\overset{k_1}{\rightleftharpoons}} E1 \tag{135}$$

$$E1 \overset{k_2}{\rightarrow} E2 + P \tag{136}$$

$$E2 \overset{k_3}{\rightarrow} E3 + Q \tag{137}$$

$$E3 \overset{k_4}{\rightarrow} E4 \tag{138}$$

$$E4 + B \underset{k_{-5}}{\overset{k_5}{\rightleftharpoons}} E5 \tag{139}$$

$$E5 \overset{k_6}{\rightarrow} E6 + R + P \tag{140}$$

$$E6 \overset{k_7}{\rightarrow} E7 + T \tag{141}$$

$$E7 + C \underset{k_{-1}}{\overset{k_1}{\rightarrow}} E8 \tag{142}$$

$$E8 \xrightarrow{k_2} E9 + P \tag{143}$$

$$E9 \xrightarrow{k_3} E10 + Q \tag{144}$$

$$E10 \xrightarrow{k_4} E11 \tag{145}$$

$$E11 + B \underset{k_{-5}}{\overset{k_5}{\rightleftharpoons}} E12 \tag{146}$$

$$E12 \xrightarrow{k_6} E13 + M + P \tag{147}$$

$$E13 \xrightarrow{k_7} E14 + T \tag{148}$$

$$E14 + D \underset{k_{-1}}{\overset{k_1}{\rightleftharpoons}} E15 \tag{149}$$

$$E15 \xrightarrow{k_2} E16 + P \tag{150}$$

$$E16 \xrightarrow{k_3} E17 + Q \tag{151}$$

$$E27 \xrightarrow{k_4} E18 \tag{152}$$

$$E18 + B \underset{k_{-5}}{\overset{k_5}{\rightleftharpoons}} E19 \tag{153}$$

$$E19 \xrightarrow{k_6} E20 + O + P \tag{154}$$

$$E20 \xrightarrow{k_7} E + T\,. \tag{155}$$

Enzyme conformations are denoted by symbols $E, E1, \ldots, E20$. Through generalization, the reaction rate covering n_c elongation cycles is expressed by:

$$V_{\mathrm{TLE},j} = \frac{[E]_t}{D} \tag{156}$$

with

$$D = \frac{1}{k_2} + \frac{1}{k_3} + \frac{1}{k_4} + \frac{1}{k_6} + \frac{1}{k_7} + \frac{k_{-5} + k_6}{k_5 k_6 [B]} + \frac{k_{-1} + k_2}{n_c k_1 k_2}\left(\frac{1}{[A]} + \frac{1}{[C]} + \frac{1}{D}\right).$$

Considering

$$k_{\mathrm{TLE},j} = \left(\frac{1}{k_2} + \frac{1}{k_3} + \frac{1}{k_4} + \frac{1}{k_6} + \frac{1}{k_7}\right)^{-1} \tag{157}$$

$$K_{\mathrm{M,T3}_j} = k_{\mathrm{TLE},j}\left(\frac{k_6 + k_{-5}}{k_5}\right) \tag{158}$$

$$K_{\mathrm{M,EFG\cdot GTP}} = \frac{k_{\mathrm{TLE},j}}{n_c}\left(\frac{k_2 + k_{-1}}{k_1}\right) \tag{159}$$

yields Eq. 74 (for $n_c \geq 1$), and Eq. 58 for the particular case where $n_c = 1$.

C
Dynamic Model of Prokaryotic Cell-Free Protein Biosynthesis

The following conditions were applied in our simulations of the cell-free synthesis of GFP.

C.1
Kinetic Model Constants

Table 10 Parameter values for the combined model for cell-free protein synthesis

Parameter	Unit	Value	Source
Transcription			
V_{T7RNAP}^{max}	μM/min	0.09	This study
$K_{M,ATP}$	μM	76	dto.
$K_{M,CTP}$	μM	34	dto.
$K_{M,GTP}$	μM	76	dto.
$K_{M,UTP}$	μM	33	dto.
$K_{M,DNA}$	μM	6.3×10^{-3}	dto.
$K_{i,GTP}$	μM	0.025	[145]
n	–	1071	This study
f_A	–	0.2652	This study
f_C	–	0.2176	dto.
f_G	–	0.2306	dto.
f_U	–	0.2866	dto.
NTPase activity			
$k_{d,NTP}$	s^{-1}	6.7×10^{-4}	This study
mRNA degradation			
$k_{D,ass}$	s^{-1}	2×10^{-4}	This study
$k_{D,Term}$	s^{-1}	50	dto.
$k_{D,endo}$	S^{-1}	2.6	dto.
$k_{D,exo}$	Nt s^{-1}	680	dto.
$k_{D,mv}$	Nt s^{-1}	95	dto.
70S initation complex formation			
$k_{TLI,70SIC}$	S^{-1}	2.5×10^{-3}	This study
$K_{M,50S}$	μM	0.011	dto.
$K_{M,fMet-tRNA_f^M}$	μM	0.053	[100]
$K_{M,mRNA}$	μM	0.01	dto.
IF2-dependent GTP hydrolysis			
$k_{TLI,IF2D}$	S^{-1}	0.8	[68]

continued on next page

Table 10 (continued)

Parameter	Unit	Value	Source
Translation elongation			
$k_{\text{TLE},j}$	s^{-1}	24	This study
$K_{\text{M,T3}_j}$	μM	0.4	dto.
$K_{\text{M,EFG}\cdot\text{GTP}}$	μM	0.22	dto.
EFG regeneration			
$k_{\text{EFG}\cdot\text{GTP}}$	$M^{-1}s^{-1}$	1.0×10^7	[110]
$k_{-\text{EFG}\cdot\text{GTP}}$	s^{-1}	400	dto.
$k_{\text{EFG}\cdot\text{GDP}}$	$M^{-1}s^{-1}$	2.7×10^7	dto.
$k_{-\text{EFG}\cdot\text{GDP}}$	s^{-1}	100	dto.
Translation termination			
k_{TLT}	s^{-1}	24	This study
$K_{\text{M,GTP}}$	μM	100	dto.
K_{M,R_K}	μM	8.3×10^{-3}	[121]
Ternary complex formation			
k_{T3_j}	$M^{-1}s^{-1}$	5×10^7	[110]
$k_{-\text{T3}_j}$	s^{-1}	1.0	dto.
tRNA charging			
$V_{\text{ARS}}^{\text{max}}$	μM/min	10	This study
$K_{\text{M,ATP}}$	μM/min	100	dto.
$K_{\text{M,aa}_j}$	μM/min	20	dto.
$K_{\text{M,tRNA}_j}$	μM/min	0.5	dto.
EFTu regeneration			
k_f	s^{-1}	30	[119]
k_r	s^{-1}	10	dto.
k_{eq}	–	0.4	This study
$K_{\text{M,EFTu}\cdot\text{GTP}}$	μM	1.0	dto.
$K_{\text{M,EFTu}\cdot\text{GDP}}$	μM	2.5	[119]
$K_{\text{M,GDP}}$	μM	3.0	[106]
$K_{i,\text{EFTu}\cdot\text{GTP}}$	μM	1.0	This study
$K_{i,\text{EFTu}\cdot\text{GDP}}$	μM	5.6	dto.
Chemical hydrolysis of AcP			
$k_{\text{d,AcP}}$	s^{-1}	3.3×10^{-5}	This study

continued on next page

Table 10 (continued)

Parameter	Unit	Value	Source
Acetate kinase			
$V^{max}_{Ack,f}$	μM/min	4000	This study
$V^{max}_{Ack,r}$	μM/min	900	dto.
K_{eq}	–	114	[135]
$K_{M,AcP}$	μM	340	dto.
$K_{M,Ac}$	μM	5800	dto.
$K_{M,ATP}$	μM	20	dto.
$K_{M,ADP}$	μM	360	dto.
$K_{i,AcP}$	μM	47	dto.
$K_{i,Ac}$	μM	100 000	dto.
$K_{i,ATP}$	μM	350	dto.
$K_{i,ADP}$	μM	50	dto.
Adenylate kinase			
$V^{max}_{Adk,f}$	μM/min	80	This study
$V^{max}_{Adk,r}$	μM/min	12	dto.
$K_{M,ATP}$	μM	51	[146]
$K_{M,ADP}$	μM	92	dto.
$K_{M,AMP}$	μM	38	dto.
Inactivation kinetics			
$k_{d,TLI}$	s^{-1}	8.9×10^{-4}	This study
$k_{d,T7RNAP}$	s^{-1}	5×10^{-5}	dto.
$k_{d,EFTu}$	s^{-1}	2.3×10^{-4}	dto.
$k_{d,EFTs}$	s^{-1}	1.9×10^{-4}	dto.

C.2 Non-Kinetic Model Constants

Table 11 Non-kinetic model constants for cell-free protein synthesis

Parameter	Unit	Value	Source
f_A	–	0.2652	This study
f_C	–	0.2176	dto.
f_G	–	0.2306	dto.
f_U	–	0.2866	dto.

C.3
Initial Conditions

Table 12 Initial conditions for simulating cell-free protein synthesis

Concentration	(μM)	Concentration	(μM)
C_{Protein}	0	C_{aa_j} (for $1 \leq i \leq A$)	250
C_j^{D} (for $1 \leq j \leq J$)	0	$C_{\text{T3},j}$ (for $1 \leq k \leq T$)	0
C_j^{M} (for $1 \leq j \leq J$)	0	$C_{\text{aa}_j\text{-tRNA}^k}$ (for $1 \leq k \leq T$)	0
C_j^{R} (for $j_{\text{R0}} \leq j \leq K$)	0	$C_{\text{fMet-tRNA}_f^{\text{M}}}$	20
C_{AcP}	34 500	$C_{\text{tRNA}}^{\text{Met}}$	0.8678
C_{ATP}	2000	$C_{\text{tRNA}}^{\text{Ala1B}}$	1.0957
C_{ADP}	106	$C_{\text{tRNA}}^{\text{Ala2}}$	0.1941
C_{AMP}	8	$C_{\text{tRNA}}^{\text{Arg2}}$	1.4002
C_{GTP}	1550	$C_{\text{tRNA}}^{\text{Arg3}}$	0.1320
C_{GDP}	75	$C_{\text{tRNA}}^{\text{Asn}}$	0.3681
C_{CTP}	1000	$C_{\text{tRNA}}^{\text{Cys}}$	0.4303
C_{CDP}	50	$C_{\text{tRNA}}^{\text{Gln1}}$	0.2242
C_{CMP}	0	$C_{\text{tRNA}}^{\text{Gln2}}$	0.3025
C_{UTP}	1000	$C_{\text{tRNA}}^{\text{Glu2}}$	1.4449
C_{UDP}	50	$C_{\text{tRNA}}^{\text{Gly12}}$	0.6594
C_{UMP}	0	$C_{\text{tRNA}}^{\text{Gly3}}$	1.2607
$C_{\text{30S},t}$	1.4	$C_{\text{tRNA}}^{\text{His}}$	0.2083
$C_{\text{50S},t}$	1.4	$C_{\text{tRNA}}^{\text{Ile12}}$	1.1365
C_{DNA}	5×10^{-3}	$C_{\text{tRNA}}^{\text{Leu1}}$	1.3246
$C_{\text{EFG},t}$	1.2120	$C_{\text{tRNA}}^{\text{Leu2}}$	0.3013
$C_{\text{EFTu},t}$	1.0605	$C_{\text{tRNA}}^{\text{Leu3}}$	0.2010
$C_{\text{EFTs},t}$	0.2727	$C_{\text{tRNA}}^{\text{Leu5}}$	0.2566
$C_{\text{IF1},t}$	0.3788	$C_{\text{tRNA}}^{\text{Lys}}$	0.5545
$C_{\text{IF2},t}$	0.4545	$C_{\text{tRNA}}^{\text{Phe}}$	0.3063
$C_{\text{IF3},t}$	0.3030	$C_{\text{tRNA}}^{\text{Pro1}}$	0.2038
$C_{\text{RF},t}$	1.7574	$C_{\text{tRNA}}^{\text{Pro2}}$	0.2275
C_{30S}	0.0065	$C_{\text{tRNA}}^{\text{Pro3}}$	0.1629
C_{50S}	0.3159	$C_{\text{tRNA}}^{\text{Ser1}}$	0.4333
C_{IF1}	0.0704	$C_{\text{tRNA}}^{\text{Ser2}}$	0.0879

continued on next page

Table 12 (continued)

Concentration	(μM)	Concentration	(μM)
C_{IF2}	0.1137	C_{tRNA}^{Ser3}	0.3430
C_{IF3}	0.0132	C_{tRNA}^{Ser5}	0.2288
C_{EFG}	0.0202	C_{tRNA}^{Thr13}	0.3402
$C_{EFG\cdot GTP}$	0.7816	C_{tRNA}^{Thr2}	0.1655
$C_{EFG\cdot GDP}$	0.4102	C_{tRNA}^{Thr4}	0.2933
$C_{EFTu\cdot GTP}$	0.7135	C_{tRNA}^{Trp}	0.2605
$C_{EFTu\cdot GDP}$	0.3467	C_{tRNA}^{Tyr12}	0.5800
C_{Ac}	136 000	C_{tRNA}^{Val1}	1.0867
C_{Pi}	0	$C_{tRNA}^{Val2A2B}$	0.3941
C_{GMP}	0	C_{tRNA}^{Asp1}	0.7232

References

1. Coburn GA, Mackie GA (1999) Proc Nucleic Acid Res Mol Biol 62:55
2. Chaney WG, Morris AJ (1979) Arch Biochem Biophys 194:283
3. Ho T, Wagner G (2004) J Biomol NMR 28:357
4. Shen LX, Basilon JP, Stanton VP (1999) PNAS 96 14:7871
5. Oresic M, Shalloway D (1998) J Mol Biol 281:31
6. Gordon R (1969) J Theor Biol 22:515
7. Vassart G, Dumont JE, Cantraine FRL (1971) Biochim Biophys Acta 247:471
8. Bergmann JE, Lodish HF (1979) J Biol Chem 254:11927
9. Liljenstrom H, Blomberg C (1987) J Theor Biol 129:41
10. Harley CB, Pollard JW, Stanners CP, Goldstein S (1981) J Biol Chem 256:10786
11. Menninger JR (1983) J Mol Biol 171:383
12. Liljenstrom H, von Heijne G (1987) J Theor Biol 124:43
13. Bagnoli F, Liò P (1995) J Theor Biol 173:271
14. Li K, Kisilevsky R, Wasan MT, Hammond G (1972) Biochim Biophys Acta 272:451
15. Singh UN (1969) J Theor Biol 25:444
16. Singh UN (1996) J Theor Biol 179:147
17. Carrier TA, Keasling JD (1997) J Theor Biol 189:195
18. Gouy M, Grantham R (1980) FEBS Lett 115:151
19. Lee SB, Bailey JE (1984) Biotechnol Bioeng 26:66
20. Biblia TA, Flickinger MC (1992) Biotechnol Bioeng 39:251
21. Kremling A, Gilles ED (2001) Metabolic Engineering 3:138
22. Hargrove JL, Schmidt FH (1989) Faseb J 3:2360
23. Hatzimanikatis V, Lee KH (1999) Metab Eng 1:275
24. Ledley TS, Ledley FD (1994) Hum Gene Ther 5:579
25. Aiba S, Humphrey AE, Millis NF (1973). Biochemical engineering. Academic Press, New York
26. Lee SB, Bailey JE (1984) Biotechnol Bioeng 26:1372

27. Chen W, Bailey JE, Lee SB (1991) Biotechnol Bioeng 38:679
28. Simha R, Zimmerman JM, Moacanin J (1963) J Chem Phys 39:1239
29. Zimmerman JM, Simha R (1965) J Theor Biol 9:156
30. Silberberg A, Simha R (1968) Biopolymers 6:479
31. Gerst I, Levine SN (1965) J Theor Biol 9:16
32. Godefroy-Colburn T, Thach RE (1981) J Biol Chem 256:11762
33. Pipkin AC, Gibbs JH (1966) Biopolymers 4:3
34. MacDonald CT, Gibbs JH, Pipkin AC (1968) Biopolymers 6:1
35. MacDonald CT, Gibbs JH (1969) Biopolymers 7:707
36. von Heijne G, Nilsson L, Blomberg C (1977) J Theor Biol 68:321
37. von Heijne G, Nilsson L, Blomberg C (1978) Eur J Biochem 92:397
38. Heinrich R, Rapaport TA (1980) J Theor Biol 86:279
39. Chela-Flores J, Liquori AM, Florio A (1988) J Theor Biol 134:319
40. Mahaffy JM (1993) J Theor Biol 162:153
41. Zhang S, Goldman E, Zubay G (1994) J Theor Biol 170:339
42. Götz P, Reuss M (1997) J Biotechnol 58:101
43. Drew DA (2001) Bull Math Biol 63:329
44. Arnold SG, Siemann M, Scharnweber K, Werner M, Baumann S, Reuss M (2001) Biotechnol Bioeng 72:548
45. Gunderson SI, Chapman KA, Burgess RR (1987) Biochemistry 26:1539
46. Blank A, Gallant JA, Burgess RR, Loeb LA (1986) Biochemistry 25:5920
47. Guajardo R, Lopez P, Dreyfus M, Sousa R (1998) J Mol Biol 281:777
48. Kolchanov NA, Ananko EA, Podkolodnaya OA, Ignatieva EV, Stepanenko IL, Kel-Margoulis OV, Kel AE, Merkulova TI, Goryachkovskaya TN, Busygina TV, Kolpakov FA, Podkolodny NL, Naumochkin AN, Romashchenko AG (1999) Nucl Acids Res 27:303
49. Wingender E, Chen X, Hehl R, Karas H, Liebich I, Matys V, Meinhardt T, Prü M, Reuter I, Schacherer F (2000) Nucl Acids Res 28:316
50. Sousa R (1996) Trends Biochem Sci 21:186
51. Blundell M, Craig E, Kennell D (1972) Nature New Biol 238:46
52. Petersen C (1993) In: Belasco JG, Brawerman G (eds) Control of messenger RNA stability. Academic, San Diego, CA, p 117
53. Court D (1993) In: Belasco JG, Brawerman G (eds) Control of messenger RNA stability. Academic, San Diego, CA, p 117
54. Rauhut R, Klug G (1999) FEMS Microbiol Rev 23:353
55. Carpousis AJ, Van Houwe G, Ehretsmann C, Krisch HM (1994) Cell 76:889
56. Py B, Higgins CF, Krisch HM, Carpousis AJ (1996) Nature 381:169
57. Miczak A, Kaberdin VR, Wei CL, Lin-Chao S (1996) Proc Natl Acad Sci USA
58. McDowall KJ, Lin-Chao S, Cohen SN (1994) J Biol Chem 269:10790
59. Bouvet P, Belasco JG (1992) Nature 360:488
60. Regnier P, Arraiano CM (2000) BioEssays 22:235
61. Belasco J, Higgins C (1988) Gene 72:15
62. Wagner LA, Gesteland RF, Dayhuff TJ, Weiss RB (1994) J Bacteriol 176:1683
63. Morse, DE, Guertin M (1971) Nature New Biol 232:165
64. Kennell D, Simmons C (1972) J Mol Biol 70:451
65. Lopez P J, Marchand I, Yarchuk O, Dreyfus M (1998) Proc Natl Acad Sci USA 95:6067
66. Rigney DR (1979) J Theor Biol 79:247
67. Lim LW, Kennell D (1979) J Mol Biol 135:369
68. Liang S-T, Ehrenberg M, Dennis P, Bremer H (1999) J Mol Biol 288:521
69. Kennell D, Riezman H (1977) J Mol Biol 114:1

70. Kennell DE (1990) In: Reznikoff UW, Gold L (eds) Maximizing gene expression. Butterworths, Boston, MA, p 101
71. Cannistraro VJ, Subbarao MN, Kennell D (1986) J Mol Biol 192:257
72. Schulz VP, Reznikoff WS (1990) J Mol Biol 211:427
73. McCormick JR, Zengel JM, Lindahl L (1991) Nucl Acids Res 19:2767
74. Schneider E, Blundell M, Kennell D (1978) Mol Gen Genet 160:121
75. Cannistraro VJ, Kennell D (1985) J Mol Biol 182:241
76. Subbarao, MN, Kennell D (1988) J Bacteriol 170:2860
77. Yarchuk O, Iost I, Dreyfus M (1991) Biochimie 73:1533
78. Liou G-G, Jane, W-N, Cohen SN, Lin N-S, Lin-Chao S (2001) Proc Natl Acad Sci USA 98:63
79. Gouy M, Gautier C (1982) Nucl Acids Res 10:7055
80. Ikemura T (1981) J Mol Biol 151:389
81. Pedersen S (1984) EMBO J 3:2895
82. Liljenstrom H, von Heijne G (1987) J Theor Biol 124:43–55
83. Sørensen MA, Pedersen S (1991) J Mol Biol 222:265
84. Varenne S, Buc J, Lloubes R, Lazdunski C (1984) J Mol Biol 180:549
85. Wolin SL, Walter P (1988) EMBO J 7:3559
86. Dahlberg AE, Lund E, Kjeldgaard NO (1973) J Mol Biol 78:627
87. Spirin AS, Lishnevskaya EB (1971) FEBS Lett 14:114
88. Naaktgeboren N, Roobol K, Voorma HO (1977) Eur J Biochem 72:49
89. Chaires JB, Pande C, Wishnia A (1981) J Biol Chem 256:6600
90. Weiel J, Hershey JWB (1982) J Biol Chem 257:1215
91. Goss DJ, Parkhurst LJ, Wahba AJ (1982) J Biol Chem 257:10119
92. Zucker FH, Hershey JWB (1986) 25:3682
93. Gualerzi C, Pon CL (1990) Biochemistry 29:5881
94. Ellis S, Conway TW (1984) J Biol Chem 259:7607
95. Wintermeyer W, Gualerzi C (1983) Biochemistry 22:690
96. Tomsic J, Vitali LA, Daviter T, Savelsbergh A, Spurio R, Striebeck P, Wintermeyer W, Rodnina M, Gualerzi CO (2000) EMBO J 19:2127
97. Canonaco MA, Calogero RA, Gualerzi CO (1986) J Mol Biol 192:257
98. Pon CL, Paci M, Pawlik RT, Gualerzi CO (1985) J Biol Chem 260:8918
99. Blumberg BM, Nakamoto T, Kezdy FJ (1979) Proc Natl Acad Sci USA 76:251
100. Gualerzi C, Risuleo G, Pon CL (1977) Biochemistry 16:1684
101. Bremer H, Dennis PP (1996) In: Neidhardt FC, Curtiss III R, Ingraham JL, Lin ECC, Brooks Low K, Magasanik B, Reznikoff WS, Riley M, Schaechter M, Umbarger HE (eds) Escherichia coli and Salmonella typhimurium, Cellular and molecular microbiology. American Society for Microbiology, Washington DC, p 1553
102. Jakubowski H (1988) J Theor Biol 133:363
103. de Smit MH, van Duin J (1994) J Mol Biol 244:144
104. Nierhaus KH (1996) Angew Chem 108:2342
105. Rohrbach MS, Bodley JW (1976) Biochemistry 15:4565
106. Hwang YW, Miller DL (1985) J Biol Chem 21:11498
107. Airas RK (1990) Eur J Biochem 192:401
108. Airas RK (1992) Eur J Biochem 210:443
109. Pavlov MY, Ehrenberg M (1996) Arch Biochem Biophys 328:9
110. Gast F-U (1987) Mechanistische Untersuchungen zur Fehlerkorrektur bei der ribosomalen Proteinsynthese. PhD thesis, University of Hannover, Germany
111. Pingoud A, Gast F-U, Peters F (1990) Biochim Biophys Acta 1050:252
112. Saifullin SR, Potapov AP (1995) Mol Biol (Mosk) 29:421

113. Saifullin SR, Potapov AP (1995) Mol Biol (Mosk) 29:434
114. Pape T, Wintermeyer W, Rodnina MV (1998) EMBO J 17:7490
115. Pingoud A, Urbanke C, Krauss G, Peters F, Maas G (1977) Eur J Biochem 78:403
116. Romero G, Chau V, Biltonen RI (1985) J Biol Chem 260:6167
117. Dong H, Nilsson I, Kurland CG (1996) J Mol Biol 260:649
118. Solomovici J, Lesnik T, Reiss C (1997) J Theor Biol 185:511
119. Ruusala T, Ehrenberg M, Kurland CG (1982) EMBO J 1:75
120. Pavlov MY, Freistroffer DV, MacDougall J, Buckingham RH, Ehrenberg M (1997) EMBO J 16:4134
121. Freistroffer DV, Pavlov MY, MacDougall J, Buckingham RH, Ehrenberg M (1997) EMBO J 16:4126
122. Voet D, Voet JG (1994) Biochemie. VCH Verlags-GmbH, Weinheim, Germany
123. Hirshfield IN, Yeh F-M (1976) Biochim Biophys Acta 435:306
124. Schulman LH, Pelka H (1988) Science 242:765
125. Schulman LH (1991) Prog Nucleic Acid Re 41:23
126. Noren CJ, Anthony-Cahill SJ, Griffith MC, Schultz PG (1989) Science 244:182
127. Hanes J, Plückthun A (1997) Proc Natl Acad Sci USA 94:4937
128. Zubay G (1973) Annu Rev Genet 7:267
129. Pratt JM (1984) In: Hames BD, Higgins SJ (eds) Transcription and translation: a practical approach. IRL, Oxford, p 179
130. Kim DM, Choi TK, Yokoyama S (1996) Eur J Biochem 239:881
131. Patnaik R, Swartz J (1998) Biotechniques 24:862
132. Kigawa T, Yabuki T, Yoshida Y, Tsutsui M, Ito Y, Shibata T, Yokoyama CS (1999) FEBS Lett 442:15
133. Chekulayeva MN, Kurnasov OV, Shirokov VA, Spirin AS (2001) Biochem Biophys Res Commun 280:914
134. Golomb M, Chamberlin M (1974) J Biol Chem 249:2858
135. Janson CA, Cleland WW (1974) J Biol Chem 249:2567
136. Reich JG, Selkov EE (1981). Energy metabolism of the cell: a theoretical treatise. Academic, London
137. Oestreich CH, Jones MM (1966) Biochemistry 5:2926
138. Schindler P, Baumann S, Siemann M, Reuss M (1999) BioTech Int J 11:12
139. Oelschlaeger P, Lange S, Schmitt J, Siemann M, Reuss M, Schmid RD (2003) Appl Microbiol Biotechnol 61:123
140. Geigenmüller U, Nierhaus KH (1990) EMBO J 9:4527
141. Katanaev VS, Spirin S, Reuss M, Siemann M (1996) FEBS Lett 397:54
142. Sambrook J, Fritsch EF, Maniatis T (1989) Molecular cloning: a laboratory manual, 2nd edn. Cold Spring Harbor Laboratory, Cold Spring Harbor, NY
143. Mailinger W, Baumeister A, Reuss M, Rizzi M (1998) J Biotechnol 63:155
144. Lippmann F, Tuttle LT (1945) J Biol Chem 159:2193:3865
145. Sen R, Dasguta D (1993) Biochem Biophys Res Commun 195:616
146. Rose T, Brune M, Wittinghofer A, Le Blay K, Surewics WK, Mantsch HH, Barzu O, Gilles AM (1991) J Biol Chem 266:10781
147. Mauch K, Arnold S, Posten C, Reuss M (1997) Computer algebra systems in model-building and model-analysis for bioprocesses. 15th IMACS World Congress 2:171–178
148. Schmid LW (1999) Reaktionskinetische Modellierung der prokaryotischen *in vitro* Translation. Studienarbeit am Institut für Bioverfahrenstechnik, Universität Stuttgart

Adv Biochem Engin/Biotechnol (2005) 100: 181–203
DOI 10.1007/b136413

Published online: 5 July 2005

Trends and Challenges in Enzyme Technology

Uwe T. Bornscheuer

Department of Technical Chemistry and Biotechnology, Institute of Chemistry and Biochemistry, Greifswald University, Soldmannstr. 16, 17487 Greifswald, Germany
uwe.bornscheuer@uni-greifswald.de

Abstract Several major developments took place in the field of biocatalysis over the past few years. These include the invention of directed evolution as an extremely useful method for biocatalyst improvement on the molecular level in combination with high-throughput screening systems, methods for accessing "nonculturable" biodiversity using metagenome approaches and progress in sequence-based biocatalyst discovery. In addition, new carriers and tools for immobilization of enzymes have been developed. For the synthesis of optically active compounds impressive examples using new enzymes and major progress in dynamic kinetic resolutions of racemates took place. These achievements are summarized in this review.

Keywords Biocatalysis · Directed evolution · Immobilization · Biodiversity · Metagenome · High-throughput screening · Dynamic kinetic resolution

Abbreviations

CLEC	Cross-linked enzyme crystals
CLEA	Cross-linked enzyme aggregates
DKR	Dynamic kinetic resolution
DMF	Dimethylformamide
E	Enantioselectivity/enantiomeric ratio

ee	Enantiomeric excess
epPCR	Error-prone PCR
FACS	Fluorescence-activated cell sorting
GC	Gas chromatography
ITCHY	Incremental truncation of chimeric hybrid enzymes
IVC	In vitro compartmentalization
StEP	Staggered extension process

1 Introduction

Biocatalysis allows the mild and selective formation of products using (mostly) isolated enzymes. Of special interest compared with chemical methods is the often observed excellent chemo-, regio- and especially stereoselectivity of biocatalysts. In the past few decades, a considerable number of processes have been developed in academia and industrialized on a commercial scale. Many examples are summarized in books [1–7] and recent reviews [8–10].

The successful development and implementation of a novel biocatalytic process requires at minimum (1) the availability of a suitable biocatalyst, (2) methods for enzyme stabilization to ease its application and re-use and (3) process engineering to deal with the choice of an appropriate reaction system (aqueous or solvent system, batch or continuous, packed-bed or membrane reactor, etc.) and with upstream and downstream processing.

Very often, a range of different enzymes can in principle be used to synthesize a certain target product. For instance, chiral alcohols can be obtained via asymmetric reduction using ketoreductases (or alcohol dehydrogenases), by kinetic resolution using lipases or esterases, or with lyases to name just a few possible enzymes. Consequently, decisions have to be made as to which enzymatic route is the best. This depends on the availability of the enzyme, its properties (specific activity, stability, pH and temperature profiles, etc.) and also price. In addition, the optical purity and isolated yield of the chiral product, the cost of starting materials and the costs for downstream processing have to be considered.

In many cases, technologies to address these requirements are readily available and only have to be adapted to the process boundaries. However, it is obvious that the availability of a suitable enzymes for a given reaction is the major precondition. Novel approaches to access biodiversity and to improve enzymes by molecular biology techniques will be addressed here. Proper stabilization of biocatalysts by advanced immobilization methods is another issue, for which novel protocols and carriers have been described and which will also be summarized in this article. Furthermore, selected examples of novel biocatalysts and dynamic kinetic resolutions (DKRs) are mentioned.

2
Accessing Biodiversity

The traditional method to identify new enzymes is based on screening of, for example, soil samples or strain collections by enrichment culture for which many impressive examples can be found in the literature [11, 12], and general references are cited in the "Introduction". Once a suitable biocatalyst is identified, strain improvement as well as cloning and expression of the encoding gene enable production on a large scale. Unfortunately, only a tiny fraction of the biodiversity can be accessed by this means using common cultivation technology. Indeed, the number of culturable microorganisms from a sample is estimated to be 0.001–1% depending on their origin [13, 14]. In turn, more than 99% of the biodiversity escaped our efforts to identify them for biocatalytic applications.

More recently, new strategies have been developed to include the plethora of "nonculturable" biodiversity in biocatalysis: (1) the metagenome approach and (2) sequence-based discovery.

Basically, in the metagenome approach, the entire genomic DNA from uncultivated microbial consortia (i.e., soil samples) is directly extracted, cloned and expressed. Microbial cells are lysed to yield high molecular weight DNA, which is then purified followed by standard cloning procedures. After propagation the DNA is usually expressed in easily cultivable surrogate host cells like *Escherichia coli*. These are then subjected to screening or selection procedures to identify distinct enzymatic activities [15–19]. The major advantage of this approach is not only that huge numbers of new biocatalysts can be found. Phylogenetic analyses revealed that new subclasses of enzymes can be identified, which show a very broad evolutionary diversity and thus the chance to identify biocatalysts with unique properties is substantially increased. In addition, the enzymes identified are already recombinantly expressed and thus in principle available on a large scale. The disadvantages are that logically only those biocatalysts can be found which can be expressed in the host organism and do not escape the activity tests.

One impressive example is the discovery of more than 130 novel nitrilases from more than 600 biotope-specific environmental DNA libraries [20], compared with fewer than 20 nitrilases known so far which were isolated by classical cultivation methods. The application of these novel nitrilases in biocatalysis revealed that 27 enzymes afforded mandelic acid in more than 90% enantiomeric excess (ee) in a DKR and one nitrilase afforded (*R*)-mandelic acid in 86% yield and 98% ee. Also, aryllactic acid derivatives were accepted at high conversion and selectivity. The best enzyme gave 98% yield and 95% ee for the (*R*) product [21] and 22 enzymes gave the opposite enantiomer with 90–98% ee. The most effective (*R*)-nitrilase was later optimized by directed evolution to withstand high substrate concentrations while maintaining high enantioselectivity [22].

Sequence-based discovery is increasingly attractive with the tremendously growing knowledge base (for lipases, epoxide hydrolases and dehalogenases, see, for example, http://www.led.uni-stuttgart.de) built from sequencing singular genes, whole genomes and even biotopes. Once new sequences are found, the cloning of the encoding genes is straightforward either by a PCR-based approach amplifying known open reading frames or by the introduction of necessary mutations in already cloned homologous enzyme genes.

3 Creating Improved Biocatalysts

Most applications of enzymes in biocatalysis do not rely on the natural reaction catalyzed by them, but rather use nonnatural substrates. In addition, the reaction system (i.e., solvent, molarity, pH, temperature) can differ substantially from the environment in which the enzymes have evolved in nature. Thus, quite often activity, stability, substrate specificity and enantioselectivity need to be improved. Until recently, these limitations were usually overcome by rather classical reaction engineering which includes variation of the reaction system until conditions are found in which the biocatalyst meets the process requirements. Nowadays, the genes encoding the biocatalyst of interest are cloned and expressed recombinantly. Consequently, variation of the enzyme by changing its amino acid sequence provides another alternative to improve its performance. In principle, two major strategies can be followed: (1) rational protein design, which requires the availability of the three-dimensional structure (or a homology model) necessary to identify type and position for the introduction of appropriate amino acid changes by site-directed mutagenesis or (2) directed evolution.

Directed evolution (also called in vitro or molecular evolution) emerged in the mid-1990s and is essentially composed of two steps: first, random mutagenesis of the gene(s) encoding the enzyme(s) and, second, identification of desired biocatalyst variants within these mutant libraries by screening or selection.

3.1 Directed Evolution

Prerequisites for in vitro evolution are the availability of the gene(s) encoding the enzyme(s) of interest, a suitable (usually microbial) expression system, an effective method to create mutant libraries and a suitable screening or selection system. Many detailed protocols for this are available from books [23–26] and reviews [27–30].

3.1.1 Methods to Create Mutant Libraries

A broad range of methods have been developed to create mutant libraries. These can be divided into two approaches, either a nonrecombining mutagenesis, in which one parent gene is subjected to random mutagenesis leading to variants with point mutations, or recombining methods in which several parental genes (usually showing high sequence homology) are randomized. This results in a library of chimeras rather than accumulation of point mutations.

One challenge in directed evolution experiments is the coverage of a sufficiently large sequence space, i.e., the creation of as many variants as possible. Considering a protein (enzyme) consisting of 200 amino acids, the number of possible variants of a protein by introduction of M substitutions in N amino acids can be calculated with the formula $19^M[N!/(N-M)!M!]$. Thus, for two random mutations already more than seven million variants are possible; with three or more substitutions, the creation and screening of a library becomes very challenging (Table 1).

The most prominent method for the creation of libraries is the error-prone PCR (epPCR) in which conditions are used which lead to the introduction of approximately one mutation per 1000 base pairs [31]. This is achieved by changing the reaction conditions, i.e., use of Mn^{2+} salts instead of Mg^{2+} salts (the polymerase is magnesium-dependent), use of the *Taq* polymerase from *Thermomyces aquaticus*, and variations in the concentrations of the desoxynucleotides. Another approach utilizes mutator strains, e.g., the *Escherichia coli* derivative *Epicurian coli* XL1-Red, lacking DNA repair mechanisms [32, 33]. Introduction of a plasmid bearing the gene encoding the protein of interest leads to mutations during replication. Both methods introduce point mutations and several iterative rounds of mutation followed by identi-

Table 1 Sequence space of possible variants for a protein consisting of 200 amino acids at a given number of substitutions

Substitutions (M)	Number of variants (sequence length $N = 200$)
1	3800
2	7 183 900
3	9 008 610 600
4	8 429 807 368 950

fication of best variants are usually required to obtain a biocatalyst with the desired properties.

Alternatively, methods of recombination (also referred to as sexual mutagenesis) can be used. The first example was the DNA-shuffling (or gene-shuffling) developed by Stemmer [34, 35], in which DNAse degrades the gene followed by recombination of the fragments using PCR with and without primers. This process mimics natural recombination and has been proven in various examples as a very effective tool to create desired enzymes. More recently, this method was further refined and termed DNA family shuffling or molecular breeding, enabling the creation of chimeric libraries from a family of genes.

The Arnold laboratory developed several methods: The staggered extension process (StEP) is based on a modified PCR protocol using a set of primers and short reaction times for annealing and polymerization. Trun-

Table 2 Selected methods to create mutant libraries for directed evolution [28, 39]

Method	Pros	Cons	Reference
Error-prone PCR	Easy to perform, mutation rate adjustable	Only point mutations accessible	[31]
Mutator strains	Easy to perform	Entire organism/ plasmid is mutated, only point mutations accessible	[32, 33]
DNA-shuffling	Modest sequence homology sufficient, several parent genes can be used, creation of chimeras possible, useful mutations are combined, harmful ones lost	Requires sequence homology	[34, 42, 63]
StEP	Similar to DNA-shuffling, simpler, no fragment purification necessary	Requires sequence homology PCR protocol must be specifically adapted	[36]
SHIPREC	No sequence homology required	Low-diversity library, in a single round (might be repeated) limited to two parents of similar length deletions/ duplications possible	[97]
ITCHY	Similar to SHIPREC	Similar to SHIPREC	[37]
THIO-ITCHY	Similar to ITCHY, but more efficient/ easier	Similar to ITCHY	[38]
GSSM	All single amino acid substitutions are covered	Technically out of reach for most researchers	[22]
SeSaM	Complete coverage at selected sites	Sites to be saturated should be known	[98]

cated oligomers dissociate from the template and anneal randomly to different templates, leading to recombination. Several repetitions allow the formation of full-length genes [36]. Other methods are incremental truncation of chimeric hybrid enzymes (ITCHY) and related approaches [37, 38]. Table 2 provides an overview of methods; more details and comparisons of different strategies for the creation of mutant libraries can be found in reviews [28, 39].

3.1.2
Assay Systems

The major challenge in directed evolution is the identification of desired variants within the mutant libraries. Suitable assay methods should enable a fast, very accurate and targeted identification of desired biocatalysts out of libraries comprising 10^4–10^6 mutants. In principle, two different approaches can be applied: screening or selection.

3.1.2.1
Selection

Selection-based systems have been used traditionally to enrich certain microorganisms. For in vitro evolution, selection methods are less frequently used as they usually can only be applied to enzymatic reactions which occur in the metabolism in the host strain. On the other hand, selection-based systems allow a considerably higher throughput compared with screening systems (see later). Often, selection is performed as a complementation, i.e., an essential metabolite is produced only by a mutated enzyme variant. For instance, a growth assay was used to identify monomeric chorismate mutases. Libraries were screened using media lacking L-tyrosine and L-phenyl alanine [40]. In a related manner, complementation of biochemical pathways has also been used to identify mutants of an enzyme involved in tryptophan biosynthesis [41]. One of these variants also retained significant HisA activity.

Stemmer's group subjected four genes of cephalosporinases from *Enterobacter*, *Yersinia*, *Citrobacter* and *Klebsiella* species to epPCR or DNA-shuffling. Libraries from four generations (a total of 50 000 colonies) were assayed by selection on agar plates with increasing concentrations of moxalactam (a β-lactam antibiotic). Only those clones could survive which were able to hydrolyze the β-lactam antibiotic. The best variants from epPCR gave only an eightfold increased activity, but the best chimeras from multiple gene-shuffling showed 270–540-fold resistance to moxalactam [42]. Sequencing of a mutant revealed low homology compared with the parental genes and a total of 33 amino acid substitutions and seven crossovers were found. These changes would have been impossible to achieve using epPCR and single-gene-shuffling only and the work demonstrates the power of DNA-shuffling.

Mutants of an esterase from *Pseudomonas fluorescens* produced by directed evolution using the mutator strain *Epicurian coli* XL1-Red were assayed for altered substrate specificity using a selection procedure [43]. Key to the identification of improved variants acting on a sterically hindered 3-hydroxy ester – which was not hydrolyzed by the wild-type esterase – was an agar plate assay system based on pH indicators, thus leading to a change in color upon hydrolysis of the ethyl ester. Parallel assaying of replica-plated colonies on agar plates supplemented with the glycerol derivative of the 3-hydroxy ester was used to refine the identification, because only *E. coli* colonies producing active esterases had access to the carbon source glycerol, thus leading to enhanced growth and in turn larger colonies. By this strategy, a double mutant which efficiently catalyzed hydrolysis was identified.

Another method is in vitro compartmentalization (IVC), which can be extended to a selection approach. IVC is based on water-in-oil emulsions, where the water phase is dispersed in the oil phase to form microscopic aqueous compartments. Each droplet contains, on average, a single gene, and serves as an artificial cell allowing for transcription, translation and the activity of the resulting proteins to take place within the compartment. The droplet volume (approximately 5×10^{-15} l) enables a single DNA molecule to be transcribed and translated [44], as well as the detection of single enzyme molecules [45]. The high capacity of the system (more than 10^{-10} in 1 ml emulsion), the ease of preparing emulsions and their high stability over a broad range of temperatures render IVC an attractive system for enzyme high-throughput screening.

IVC provides a facile means for co-compartmentalizing genes and the proteins they encode, but the selection of an enzymatic activity requires a link between the desired reaction product and the gene (Fig. 1). One possible selection format is to have the substrate, and subsequently the product, of the desired enzymatic activity physically linked to the gene. Enzyme-encoding genes can then be isolated by virtue of their attachment to the product, while other genes that encode an inactive protein carry the unmodified substrate. The simplest applications of this strategy lies in the selection of DNA-modifying enzymes where the gene and substrate comprise the same molecule. Indeed, IVC was first applied for the selection of DNA-methyltransferases (MTases) [44]. Selection was performed by extracting the genes from the emulsion and subjecting them to digestion by a cognate restriction enzyme that cleaves the non-methylated DNA [46–48]. Other applications can be found in a recent review [49].

In addition, IVC can also be performed in double emulsions. An alternative strategy has been developed based on compartmentalizing, and sorting, single genes, together with the fluorescent product molecules generated by their

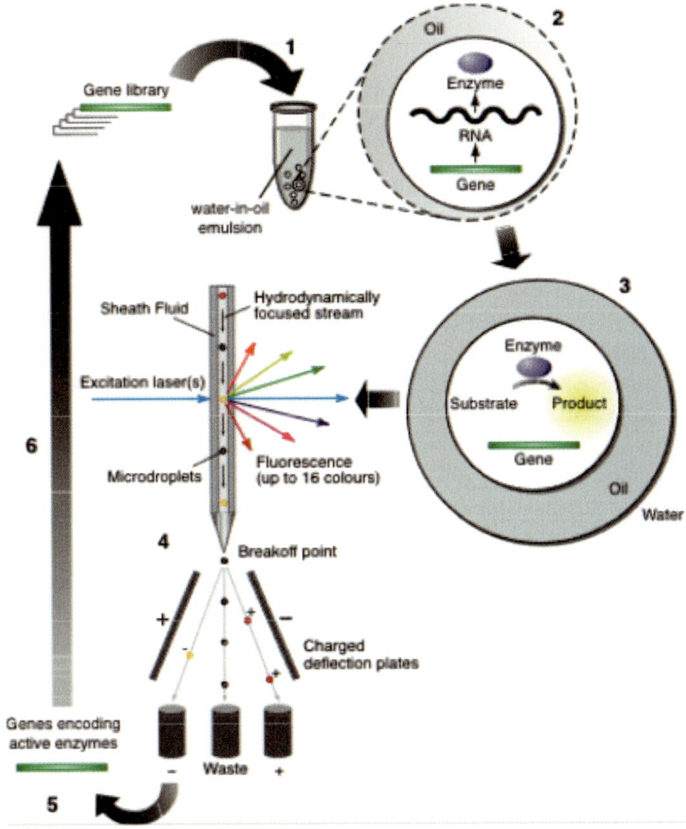

Fig. 1 Selections by flow-sorting of double emulsion microdroplets using a fluorescence-activated cell sorter (*FACS*). A library of genes, each encoding a different enzyme variant, is dispersed to form a water-in-oil (*w/o*) emulsion with typically one gene per aqueous microdroplet (*1*). The genes are transcribed and translated within their microdroplets (*2*), using either in vitro (cell-free) transcription/translation or by compartmentalizing single cells (e.g., bacteria into which the gene library is cloned) in the microdroplets. Proteins with enzymatic activity convert the nonfluorescent substrate into a fluorescent product and the w/o emulsion is converted into a water-in-oil-in-water emulsion (*3*). Fluorescent microdroplets are separated from nonfluorescent microdroplets (or microdroplets containing differently colored fluorochromes) using a FACS (*4*). Genes from fluorescent microdroplets, which encode active enzymes, are recovered and amplified (*5*). These genes can be recompartmentalized for further rounds of selection (*6*)

encoded enzymes. The technology makes use of double water-in-oil-in-water emulsions that are amenable to sorting by fluorescence-activated cell sorting (FACS). It circumvents the need to tailor the selection for each substrate and reaction, and allows the use of a wide variety of existing fluorogenic substrates [50].

3.1.2.2 Screening

Much more frequently used are screening-based systems (not to be confused with the use of the term "screening" for the identification of microorganisms). Owing to the very high number of variants generated by directed evolution, common analytical tools like gas chromatography (GC) and high-performance liquid chromatography are less useful, as they are usually too time-consuming. Also high-throughput GC–mass spectrometry or NMR techniques have been described, but these require the availability of rather expensive equipment and in the case of screening for enantioselective biocatalysts also the use of deuterated substrates. In addition, phage display [52], ribosome display and FACS have been used to screen within mutant libraries. Although they allow the screening of mutant libraries on the order of $> 10^6$ variants, they are hardly generally applicable.

The most frequently used methods are based on photometric and fluorimetric assays performed in microtiter-plate-based formats in combination with high-throughput robot assistance. They allow a rather accurate screening of several tens of thousands of variants within a reasonable time and provide sufficient information about the enzymes investigated, i.e., the activity by determining the initial rates or endpoints and stereoselectivity by using both enantiomers of the compound of interest. One versatile example is the use of umbelliferone derivatives (Scheme 1). Esters or amides of umbelliferone are rather unstable, especially at extreme pH and at elevated temperatures. The ether derivatives shown in Scheme 1 are very stable as the fluorophore is linked to the substrate via an ether bond. Only after enzymatic reaction and treatment with sodium periodate and bovine serum albumin is the fluorophore released [51].

Another alternative is the recently described "surface-enhanced resonance Raman scattering", which was shown to enable a rapid and highly sensi-

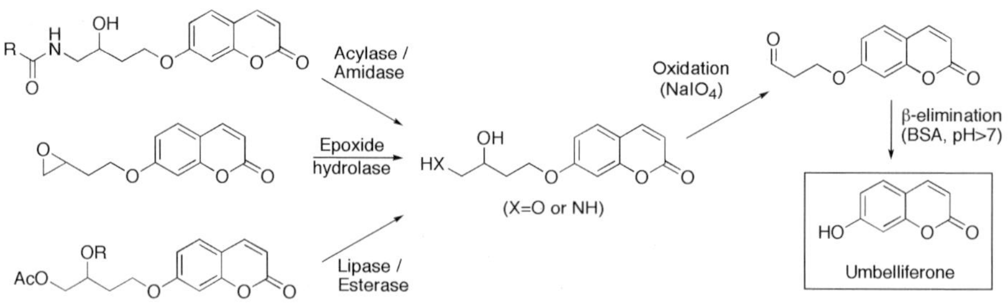

Scheme 1 Fluorogenic assay based on umbelliferone derivatives. Enzyme activity yields a product which upon oxidation with sodium periodate and treatment with bovine serum albumin (*BSA*) yields umbelliferone [51]

tive identification of lipase activity and enantioselectivity on dispersed silver nanoparticles [53, 54].

A variety of further assay methods can be found in a number of recent reviews [55–58].

3.1.3
Examples

Reetz and coworkers turned a nonenantioselective (2% ee $E = 1.1$) lipase from *Ps. aeruginosa* PAO1 into a variant with very good selectivity ($E > 51$, more than 95% ee) in the kinetic resolution of 2-methyldecanoate. Identification of variants was based on optically pure (*R*)-*p*-nitrophenyl and (*S*)-*p*-nitrophenyl esters of 2-methyldecanoate in a spectrophotometric screening. In the first step, the wild-type lipase gene was subjected to several rounds of random mutagenesis by epPCR leading to a variant with $E = 11$ (81% ee) followed by saturation mutagenesis ($E = 25$). Key to further doubling of enantioselectivity was a combination of DNA-shuffling, combinatorial cassette mutagenesis and saturation mutagenesis, which led to a maximal recombination of the best variants. The best mutant ($E > 51$) contained six amino acid substitutions and a total of approximately 40 000 variants were screened [59]. The overall strategy is illustrated in Fig. 2; the overall changes in enantioselectivity using the combination of different approaches for random mutagenesis are summarized in Fig. 3.

The Arnold group reported the inversion of enantioselectivity of a hydantoinase from D-selectivity (40% ee) to moderate L-preference (20% ee at 30% conversion) by a combination of epPCR and saturation mutagenesis. Only one amino acid substitution was sufficient to invert enantioselectivity. Thus, production of L-methionine from D,L-5-(2-methylthioethyl)hydantoin in a whole-cell system of recombinant *E. coli* also containing a L-carbamoylase and a racemase at high conversion became feasible [60].

Even if a biocatalyst with proper substrate specificity (and stereoselectivity) is already identified, the requirements for a cost-effective process are not always fulfilled. Enzyme properties such as pH, temperature and solvent stability are very difficult to improve by "classical" methods like immobilization techniques or site-directed mutagenesis. Again, directed evolution has been shown to be a versatile tool to meet this challenge.

For instance, an esterase from *Bacillus subtilis* hydrolyzes the *p*-nitrobenzyl ester of loracarbef, a cephalosporin antibiotic. Unfortunately, the wild-type enzyme was only weakly active in the presence of dimethylformamide (DMF), which must be added to dissolve the substrate. A combination of epPCR and DNA-shuffling led to the generation of a variant with 150 times higher activity compared with that of the wild-type in 15% DMF [61]. Later, the thermostability of this esterase could also be increased by approximately 14 °C

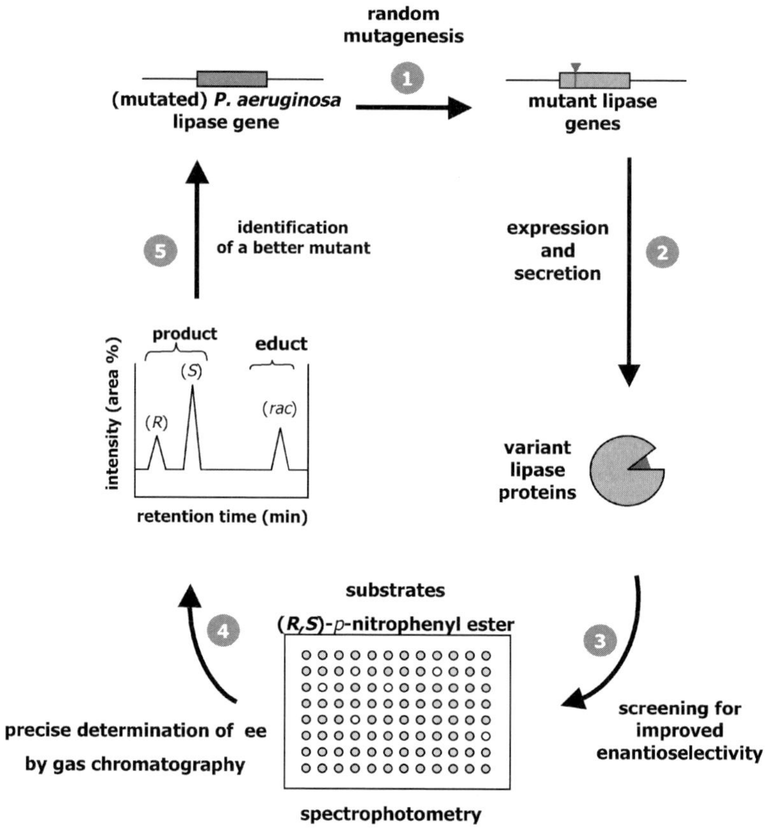

Fig. 2 Directed evolution of a lipase from *Pseudomonas aeruginosa* for the enantioselective resolution of 2-methyl decanoate. In the first step (*1*), the lipase gene was subjected to random mutagenesis, next the mutated genes were expressed and secreted (*2*). Screening for improved enantioselectivity was based on a spectrophotometric assay using optically pure (*R*)-*p*-nitrophenyl or (*S*)-*p*-nitrophenyl esters of the substrate (*3*). Hit mutants with improved enantioselectivity were then verified by gas chromatography (*4*). The cycle was repeated several times to identify the best mutants (*5*) [59]

by directed evolution. In a similar manner, the performance of subtilisin E in DMF was improved 470-fold.

It could also been shown that it is possible to increase the thermostability of a cold-adapted protease to 60 °C while maintaining high activity at 10 °C [62]. The best psychrophilic subtilisin S41 variant contained only seven amino acid substitutions resembling only a tiny fraction of the usual 30–80% sequence difference found between psychrophilic enzymes and mesophilic counterparts.

In another example, researchers at Maxygen (USA) and Novozymes (Denmark) simultaneously screened for four properties in a library of family-

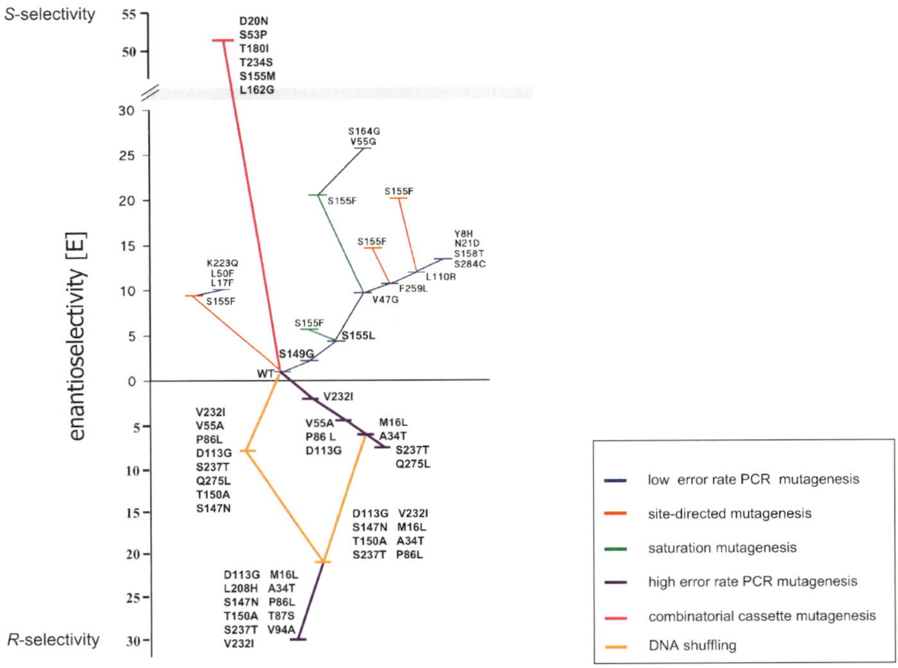

Fig. 3 Changes in enantioselectivity of a lipase from *Ps. aeruginosa* using methods of directed evolution. Starting from the nonselective wild-type ($E = 1.1$), the combination of various genetic tools led to the creation and identification of variants with high (*S*)-selectivity ($E = 51$) and with good (*R*)-selectivity ($E = 30$) [59]

shuffled subtilisins (activity at 25 °C, thermostability, organic-solvent tolerance and pH profile) and reported variants with considerably improved characteristics for all parameters [63].

4
Dynamic Kinetic Resolution vs. Asymmetric Synthesis

A kinetic resolution of a racemate can only yield at maximum 50% product. In order to achieve a complete conversion of both enantiomers, a DKR can be used. Such a strategy can also make the synthesis of an optically pure compound more competitive to an asymmetric synthesis using, e.g., alcohol dehydrogenases and a prochiral substrate (Scheme 2).

The requirements for a DKR are (1), the substrate must racemize faster than the subsequent enzymatic reaction proceeds, (2) the product must not racemize and (3) as in any asymmetric synthesis, the enzymatic reaction

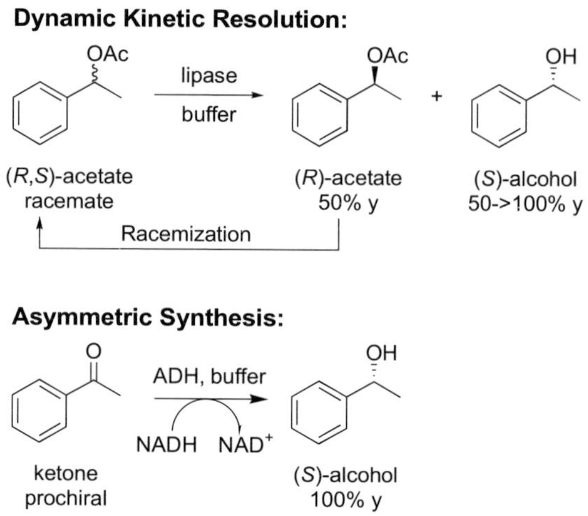

Scheme 2 A dynamic kinetic resolution of a racemic alcohol by a lipase can provide similar to an asymmetric synthesis using an alcohol dehydrogenase (*ADH*) theoretically up to 100% yield of one enantiomer in optically pure form. This requires a suitable racemization method (enzymatic or chemically)

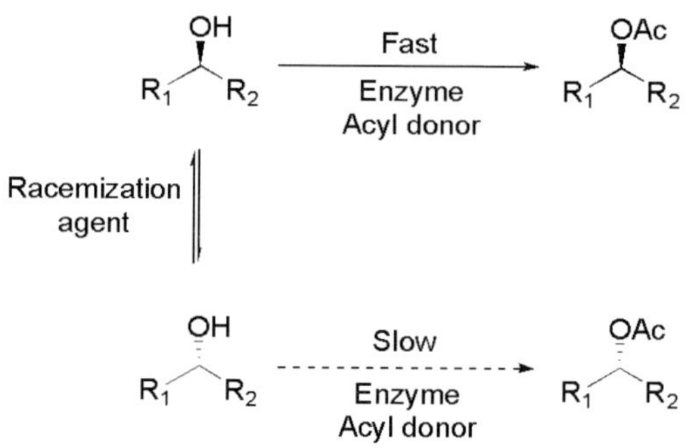

Scheme 3 Principle of a dynamic kinetic resolution

must be highly stereoselective (Scheme 3). Many examples are covered in recent reviews [64–67].

An early example of a DKR was the synthesis of optically pure α-amino acids from hydantoins, a process which is currently performed in industry using an engineered *E. coli* strain expressing all three required enzymes (hydantoinase, carbamoylase and racemase) (Scheme 4). Racemization of the hydantoin can also be performed at alkaline pH [60, 68, 69].

Later, DKRs were described for desymmetrizations of chemically labile secondary alcohols, thiols and amines (i.e., cyanohydrins, hemiacetals, hemithioacetals). More recently, in situ deracemization via nucleophilic displacement has been demonstrated for 2-chloropropionate (92% yield, 86%

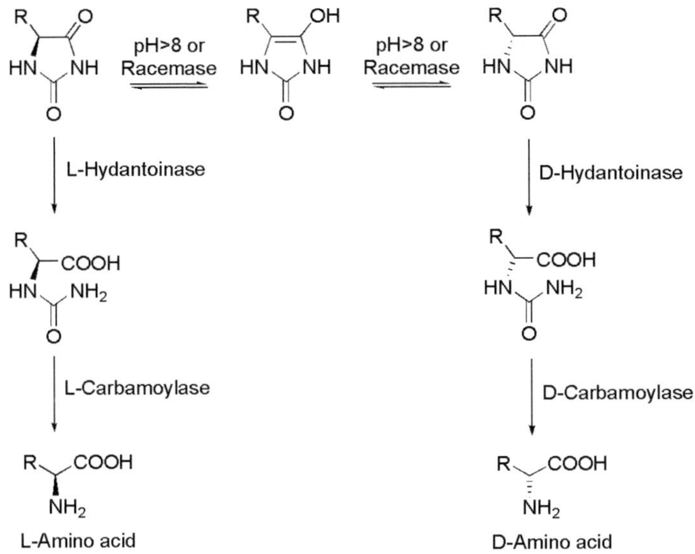

Scheme 4 Synthesis of L- or D-amino acids using a combination of hydantoinase, carbamoylase and racemase. This process can be performed using an engineered whole-cell system with an *Escherichia coli* strain

ee) using lipase from *Candida cylindracea* in an aminolysis supported by triphenylphosphonium chloride [70].

Other approaches are combinations of enzymatic resolution with metal-catalyzed racemization. They usually proceed either via hydrogen transfer or via π-allyl-complex formation. Bäckvall and coworkers developed a hydrogen transfer system based on a ruthenium catalyst with *p*-chloroethyl acetate as acyl donor. Enolesters – with the exception of isopropenyl acetate – cannot be used owing to side reactions. On the other hand, no addition of ketones or ex-

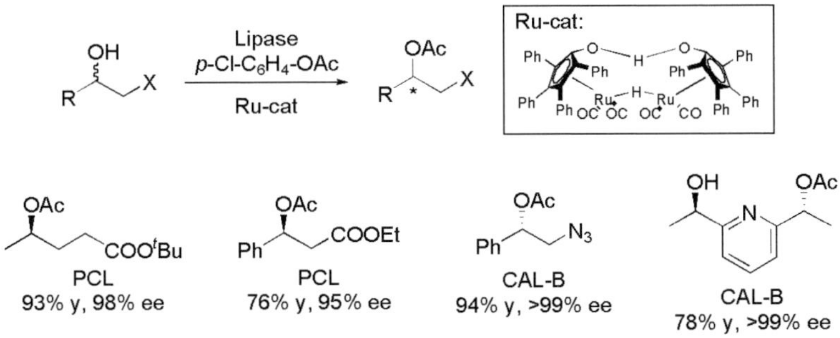

Scheme 5 Examples of the dynamic kinetic resolution of secondary alcohols using a ruthenium catalyst

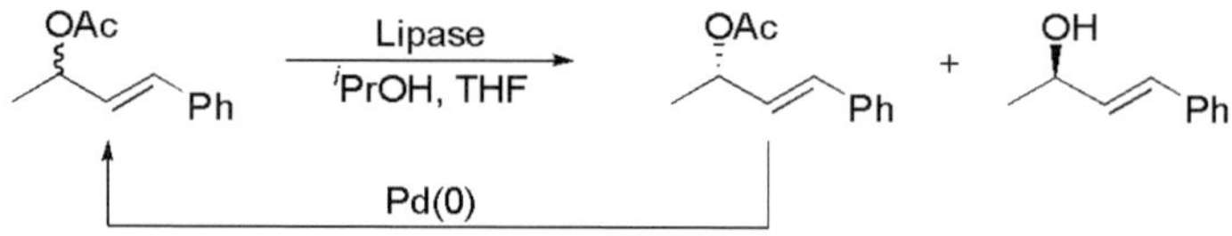

Scheme 6 Example of the dynamic kinetic resolution of an allylic alcohol using Pd(0)

ternal bases is required, which often affect the reaction performance. Selected examples are shown in Scheme 5.

Kim and coworkers improved the DKR of allylic acetates using Pd(0) catalysts in tetrahydrofuran. 2-Propanol serves as an acyl acceptor and the unreactive enantiomer is racemized by $Pd(PPh)_3$ with added diphosphine at room temperature (Scheme 6). A series of linear allylic acetates were deracemized in high ee (97–99% ee) and with moderate to good yields (61–78%).

Recently, a deracemization of α-methylbenzyl amine using a monoamine oxidase from *Aspergillus niger*in combination with a chemical nonselective reduction step using, for instance, sodium borohydride or amine borane was described (Scheme 7). Overall, this process led to the formation of optically active amines from the racemate. Directed evolution of this enzyme resulted in an amine oxidase possessing not only a wider substrate spectrum, but also good enantioselectivity. The Asn336Ser variant of the amine oxidase showed highest activity towards substrates bearing a methyl substituent and a bulky alkyl/aryl group adjacent to the amino carbon atom. In all cases examined so far, the enzyme variant was enantioselective for the (*S*)-isomer of the racemic amine substrate [71–73].

In special cases, the resolution of a racemate can lead to only one enantiomer. This includes the enantioconvergent hydrolysis of epoxides. This was achieved using two complementary epoxide hydrolases [74]. The enzyme from *A. niger* hydrolyzed one enantiomer via attack at C-2 with retention of configuration, while the epoxide hydrolase from *Beauveria sulfurescens* attacked at C-1 with inversion of configuration. Thus, a mixture of both enzymes produced the (*R*)-diol (Scheme 8).

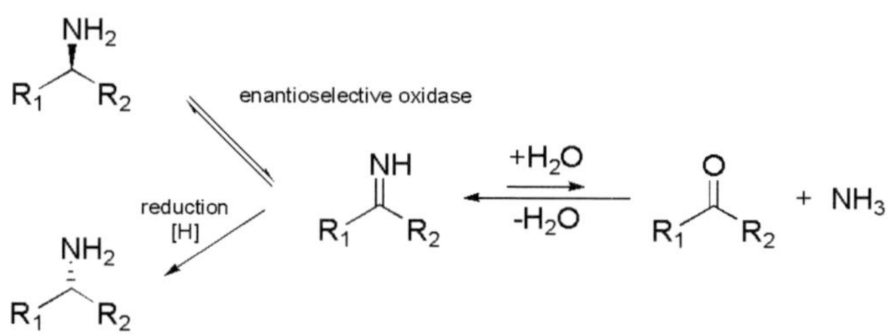

Scheme 7 The deracemization of chiral amines using a sequence of enantioselective oxidation using an amine oxidase coupled with a nonselective reducing agent

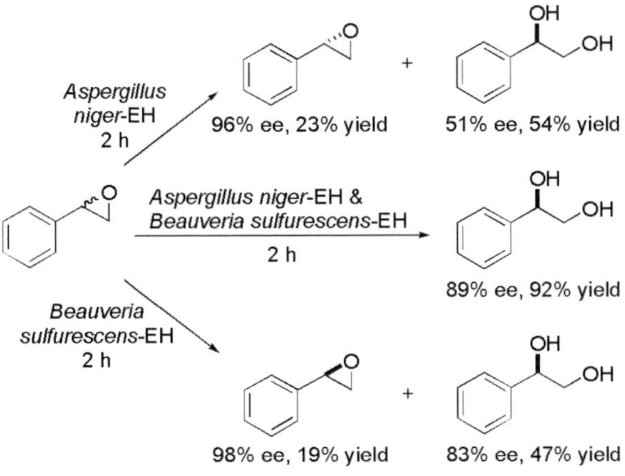

Scheme 8 Enantioconvergent kinetic resolution of an epoxide using two complementary epoxide hydrolases

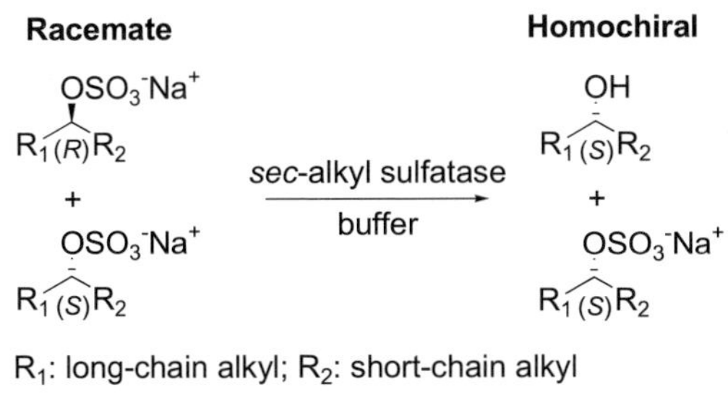

Scheme 9 A deracemization process using alkyl sulfatases can lead to homochiral products

More recently, alkyl sulfatases were discovered, which perform substrate hydrolysis via inversion and therefore enable a deracemization process too (Scheme 9). Thus, both the secondary alcohol formed as a product and the remaining unconverted sulfate ester possess the same absolute configuration and hence constitute a homochiral product mixture [75]. Unfortunately, the enantioselectivities of the *Rhodococcus* sulfatase ranged from low to moderate only ($E \leq 21$). Addition of Fe^{3+} can lead to enhanced enantioselectivities [76].

5 Other examples

In contrast to epoxide hydrolases, which do not accept nucleophiles other than water and consequently only catalyze the formation of a diol from an epoxide, haloalcohol dehalogenases (also known as halohydrin dehaloge-

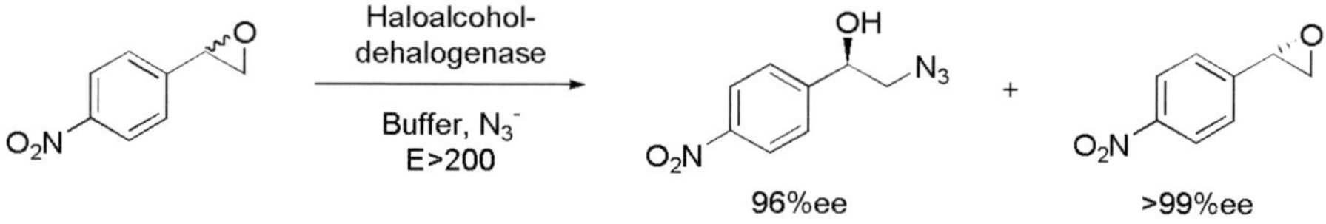

Scheme 10 A haloalcohol dehalogenase from *Agrobacterium radiobacter* also accepts an azide as a nucleophile in the highly enantioselective ring opening of an epoxide

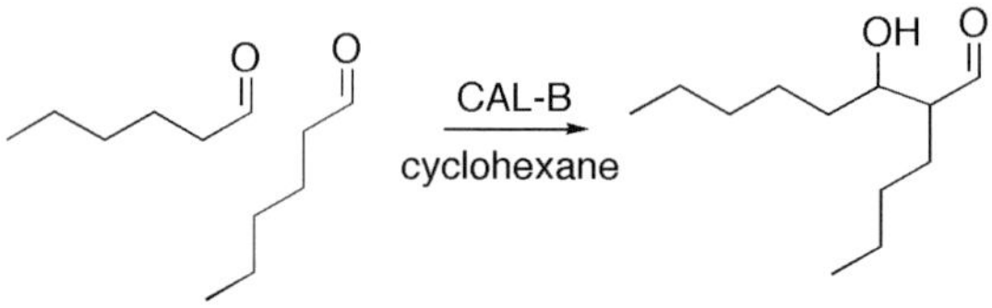

Scheme 11 Lipase B from *Candida antarctica* also catalyzed an aldol addition of hexanal, an example for catalytic promiscuity. The lyase activity is more than 10^5 times slower than the hydrolysis of a triglyceride, but still faster than aldol additions catalyzed by a catalytic antibody with aldolase activity

nases, hydrogen halide lyases and halohydrin epoxidases), also accept nucleophiles like CN^-, NO_2^- and N_3^- beside the natural nucleophile halide (Cl^-, Br^-, I^-). The resulting products are important intermediates in the synthesis of amino alcohols. An example is shown in Scheme 10 for the reaction catalyzed by a haloalcohol dehalogenase from *Agrobacterium radiobacter* [77, 78].

Over the last few years, evidence has been mounting that enzymes do not catalyze only one single chemical transformation, but are also able to perform several types of reactions. This ability is termed catalytic promiscuity and does not only exist among a few enzymes, but appears to be rather common [79–81]. Examples include single proteins with several catalytic abilities and also where small changes (typically metal ion substitutions or site-directed mutagenesis) introduce new catalytic activity. The most successful examples are carbon–carbon bond forming reactions, oxidations catalyzed by hydrolytic enzymes and glycosyl transfer reactions. For instance, it was found that lipase B from *C. antarctica* (lipases belong to enzyme class EC 3.1.1.3) is also able to catalyze a carbon–carbon bond forming reaction (an aldol addition, usually catalyzed by a lyase, EC class 4) [82] (Scheme 11). Although the reaction was not enantioselective, the diastereoselectivity differed from the spontaneous reaction. The authors hypothesized that the aldol addition did not require the active site serine and, indeed, replacement with alanine (Ser105Ala) increased the aldol addition approximately twofold.

6 Advances in Immobilization Technologies

Even if an enzyme is identified to be useful for a given reaction, its application is often hampered by its lack of long-term stability under process conditions, and also by difficulties in recovery and recycling.

This problem can be overcome by immobilization, providing advantages such as enhanced stability, repeated or continuous use, easy separation from the reaction mixture and possible modulation of catalytic properties. Since the first uses of biocatalysts in organic synthesis dating back almost a century, researchers have tried to identify methods to link an enzyme to a carrier. Numerous examples for a broad range of enzymes and reaction systems (aqueous system, organic solvents) have been documented in the literature [83, 84], which reflects the importance of biocatalysis. On the other hand this also exemplifies that a general, broadly applicable method for enzyme immobilization still needs to be discovered. The most frequently used immobilization techniques fall into four categories: (1) noncovalent adsorption or deposition, (2) covalent attachment, (3) entrapment into a polymeric gel and (4) cross-linking of an enzyme. All these approaches are a compromise between maintaining high catalytic activity while achieving the advantages given before.

Two recent trends are (1) the use of novel reagents and/or carriers and (2) approaches taking into account increasing knowledge about enzyme structure and mechanism [85].

As early as 1995, Reetz et al. [86] reported that immobilization in sol–gels can enhance the activity of lipases up to 100-fold. For cross-linked enzyme crystals (CLECs) [87, 88], an increase in enantioselectivity compared with that of the native enzyme was described [89], but this was mostly attributed to the removal of a less selective isoenzyme during CLEC preparation. As crystallization of proteins is not an easy task, cross-linked enzyme aggregates (CLEA) obtained by precipitation of proteins followed by cross-linking with glutaraldehyde might represent an easy alternative. The CLEA from pencillin acylase had the same activity as a CLEC in the synthesis of ampicillin, but a cross-linked aggregate also catalyzed the reaction in a broad range of organic solvents [90].

A promising combination of easy separation and high stability has been reported for a lipase immobilized on γ-Fe_2O_3 magnetic nanoparticles [91]. The use of magnetic particles is not new [92, 93], but Ulman and coworkers were able to produce nanoparticles with an average size of 20 ($\pm$10 nm) (usually 75–100 μm), which were then covalently linked after thiophene functionalization to a lipase from *C. rugosa*. The resulting biocatalyst exhibited significantly higher stability (over a period of almost 1 month) than the native enzyme in the hydrolysis of *p*-nitrophenylbutyrate. Moreover, separation of the immobilized enzyme from the reactant mixture by a magnetic field holding either the immobilized enzyme in place or removing it is facilitated more as the nanoparticles show very high magnetization values.

The increasing knowledge of enzyme structures and mechanism should also enable more controlled immobilizations. For example, lipase from *Ps. fluorescens* was immobilized on four different carriers [94]. The native enzyme and two carrier-linked lipase preparations show no or only modest changes in activity and enantioselectivity in the kinetic resolution of a racemic carboxylic acid ethylester. However, two immobilisates exhibited substantially altered properties. Specific activity was increased 10-fold and enantioselectivity increased from $E = 7$ to $E = 86$ for lipase immobilized on decaoctyl sepharose. The authors claim that during this (also much rapider) immobilization procedure the lipase underwent a conformational change from the closed to an open structure, as a hydrophobic "lid" – known to be present in most lipases – moves aside by an interfacial activation caused by the carrier and the immobilization procedure, providing enhanced substrate access to the active-site residues. With a similar strategy, the same group also reported modulation of the properties of penicillin acylases from three different species which also undergo conformational changes upon binding of the acyl donor substrate [95, 96].

7
Conclusions and Perspectives

The examples summarized in this review demonstrate that biocatalysis is rapidly developing and is still a growing field. Compared with the technologies used about 15–20 years ago, a substantial change can be observed. Most of all, this includes the vast developments in molecular biology tools and bioinformatics highlighted here, which have become the major driving forces in biocatalyst discovery and improvement. This is further boosted by the growing interest in biocatalysts to replace conventional chemical processes. On one hand, the new methodologies will continue to lead to the creation of better enzymes of well-known activity (e.g., lipase, esterase, nitrilase, hydantoinase); on the other hand, the discovery of new enzymes with novel properties interesting to chemists (e.g., alkyl sulfatase, haloalkohol dehalogenase) opens new alternatives in the field of white biotechnology.

Acknowledgements Financial support by the Fonds der Chemischen Industrie (Frankfurt, Germany) is gratefully acknowledged. I also thank Karl-Erich Jäger (Jülich, Germany) for the provision of Figs. 2 and 3.

References

1. Liese A, Seelbach K, Wandrey C (2000) Industrial biotransformations. Wiley-VCH, Weinheim
2. Drauz K, Waldmann H (2002) Enzyme catalysis in organic synthesis, 2nd edn, vols 1–3. VCH, Weinheim
3. Bommarius AS, Riebel BR (2004) Biocatalysis, vol 1. Wiley-VCH, Weinheim
4. Patel RN (2000) Stereoselective biocatalysis. Dekker, New York
5. Faber K (2004) Biotransformations in organic chemistry, 4th edn. Springer, Berlin Heidelberg New York
6. Bornscheuer UT, Kazlauskas RJ (1999) Hydrolases in organic synthesis – regio- and stereoselective biotransformations. Wiley-VCH, Weinheim
7. Buchholz K, Kasche V, Bornscheuer UT (2005) Biocatalysts and enzyme technology. Wiley-VCH, Weinheim
8. Schoemaker HE, Mink D, Wubbolts MG (2003) Science 299:1694
9. Schmid A, Dordick JS, Hauer B, Kiener A, Wubbolts M, Witholt B (2001) Nature 409:258
10. Breuer M, Ditrich K, Habicher T, Hauer B, Keßeler M, Stürmer R, Zelinski T (2004) Angew Chem Int Ed Engl 43:788
11. Ogawa J, Shimizu S (2002) Curr Opin Biotechnol 13:367
12. Asano Y (2002) J Biotechnol 94:65
13. Lorenz P, Liebeton K, Niehaus F, Schleper C, Eck J (2003) Biocat Biotransf 21:87
14. Miller CA (2000) Inform 11:489
15. Handelsman J (2005) Nat Biotechnol 23:38
16. Handelsman J (2004) Microbiol Mol Biol Rev 68:669
17. Lorenz P, Eck J (2004) Eng Life Sci 4:501
18. Uchiyama T, Takashi A, Ikemura T, Watanabe K (2005) Nat Biotechnol 23:88
19. Short JM (1997) Nat Biotechnol 15:1322
20. Robertson DE, Chaplin JA, DeSantis G, Podar M, Madden M, Chi E, Richardson T, Milan A, Miller M, Weiner DP, Wong K, McQuaid J, Farwell B, Preston LA, Tan X, Snead MA, Keller M, Mathur E, Kretz PL, Burk MJ, Short JM (2004) Appl Environ Microbiol 70:2429
21. DeSantis G, Zhu Z, Greenberg WA, Wong K, Chaplin J, Hanson SR, Farwell B, Nicholson LW, Rand CL, Weiner DP, Robertson DE, Burk MJ (2002) J Am Chem Soc 124:9024
22. DeSantis G, Wong K, Farwell B, Chatman K, Zhu Z, Tomlinson G, Huang H, Tan X, Bibbs L, Chen P, Kretz K, Burk MJ (2003) J Am Chem Soc 125:11476
23. Arnold FH, Georgiou G (eds) (2003) Directed enzyme evolution: screening and selection methods. Methods in molecular biology, vol 230. Humana, Totawa
24. Arnold FH, Georgiou G (eds) (2003) Directed evolution library creation: methods and protocols. Methods in molecular biology, vol 231. Humana, Totawa
25. Brakmann S, Johnsson K (2002) Directed molecular evolution of proteins, vol 1. Wiley-VCH, Weinheim, p 357
26. Brakmann S, Schwienhorst A (2004) Evolutionary methods in biotechnology: clever tricks for directed evolution. Wiley-VCH, Weinheim
27. Reetz MT (2004) Proc Natl Acad Sci USA 101:5716
28. Neylon C (2004) Nucl Acid Res 32:1448
29. Turner NJ (2003) Trends Biotechnol 21:474
30. Bornscheuer UT (2001) Biocat Biotransf 19:84
31. Cadwell RC, Joyce GF (1992) PCR Meth Appl 2:28
32. Greener A, Callahan M, Jerpseth B (1996) Methods Mol Biol 57:375

33. Bornscheuer UT, Altenbuchner J, Meyer HH (1998) Biotechnol Bioeng 58:554
34. Stemmer WPC (1994) Proc Natl Acad Sci USA 91:10747
35. Stemmer WP (1994) Nature 370:389
36. Zhao H, Giver L, Shao Z, Affholter JA, Arnold FH (1998) Nat Biotechnol 16:258
37. Ostermeier M, Nixon AE, Benkovic SJ (1999) Bioorg Med Chem 7:2139
38. Lutz S, Ostermeier M, Benkovic SJ (2001) Nucl Acids Res 29:1
39. Kurtzman AL, Govindarajan S, Vahle K, Jones JT, Heinrichs V, Patten PA (2001) Curr Opin Biotechnol 12:361
40. MacBeath G, Kast P, Hilvert D (1998) Science 279:1958
41. Juergens C, Strom A, Wegener D, Hettwer S, Wilmanns M, Sterner R (2000) Proc Natl Acad Sci USA 97:9925
42. Crameri A, Raillard SA, Bermudez E, Stemmer WP (1998) Nature 391:288
43. Bornscheuer UT, Altenbuchner J, Meyer HH (1999) Bioorg Med Chem 7:2169
44. Griffiths AD, Tawfik DS (1998) Nat Biotechnol 16:652
45. Griffiths AD, Tawfik DS (2003) EMBO J 22:24
46. Tawfik DS, Griffiths AD (1998) Nat Biotechnol 16:652
47. Lee YF, Tawfik DS, Griffiths AD (2002) Nucl Acids Res 30:4937
48. Cohen HM, Tawfik DS, Griffiths AD (2004) Protein Eng Des Sel 17:3
49. Aharoni A, Griffiths AD, Tawfik DS (2005) Curr Opin Chem Biol 9:210
50. Bernath K, Hai M, Mastrobattista E, Griffiths AD, Magdassi S, Tawfik DS (2004) Anal Biochem 325:151
51. Reymond JL, Wahler D (2002) Chem Bio Chem 3:701
52. Fernandez-Gacio A, Uguen M, Fastrez J (2003) Trends Biotechnol 21:408
53. Moore BD, Stevenson L, Watt A, Flitsch S, Turner NJ, Cassidy C, Graham D (2004) Nat Biotechnol 22:1133
54. Bornscheuer UT (2004) Nat Biotechnol 22:1098
55. Goddard JP, Reymond J-L (2004) Trends Biotechnol 22:363
56. Bornscheuer UT (2001) Biocat Biotransf 19:84
57. Wahler D, Reymond JL (2001) Curr Opin Biotechnol 12:535
58. Reetz MT (2002) Angew Chem Int Ed Engl 41:1335
59. Reetz MT, Wilensek S, Zha D, Jaeger K-E (2001) Angew Chem Int Ed Engl 40:3589
60. May O, Nguyen PT, Arnold FH (2000) Nat Biotechnol 18:317
61. Moore JC, Arnold FH (1996) Nat Biotechnol 14:458
62. Miyazaki K, Wintrode PL, Grayling RA, Rubingh DN, Arnold FH (2000) J Mol Biol 297:1015
63. Ness JE, Welch M, Giver L, Bueno M, Cherry JR, Borchert TV, Stemmer WP, Minshull J (1999) Nat Biotechnol 17:893
64. El Gihani MT, Williams JMJ (1999) Curr Opin Biotechnol 3:11
65. Kim J-M, Ahn Y, Park J (2002) Curr Opin Biotechnol 13:578
66. Pàmies O, Bäckvall J-E (2004) Trends Biotechnol 22:130
67. Pàmies O, Bäckvall J-E (2003) Chem Rev 103:3247
68. Altenbuchner J, Siemann-Herzberg M, Syldatk C (2001) Curr Opin Biotechnol 12:559
69. Park JH, Kim GJ, Kim HS (2000) Biotechnol Prog 16:564
70. Bdjìc JD, Kadnikova EN, Kostic NM (2001) Org Lett 3:2025
71. Alexeeva M, Enright A, Dawson MJ, Mahmoudian M, Turner NJ (2002) Angew Chem Int Ed Engl 41:3177
72. Alexeeva M, Carr R, Turner NJ (2003) Org Biomol Chem 1:4133
73. Carr R, Alexeeva M, Enright A, Eve TS, Dawson MJ, Turner NJ (2003) Angew Chem Int Ed Engl 42:4807
74. Pedragosa-Moreau S, Archelas A, Furstoss R (1993) J Org Chem 58:5533

75. Pogorevc M, Kroutil W, Wallner SM, Faber K (2002) Angew Chem Int Ed Engl 41:4052
76. Pogorevc M, Strauss UT, Riermeier TH, Faber K (2002) Tetrahedron Asymmetry 13:1443
77. Spelberg JH, van Hylckama Vlieg JE, Tang L, Janssen DB, Kellogg RM (2001) Org Lett 3:41
78. Spelberg JH, Tang L, van Gelder M, Kellogg RM, Janssen DB (2002) Tetrahedron Asymmetry 13:1083
79. Bornscheuer UT, Kazlauskas RJ (2004) Angew Chem Int Ed Engl 43:6032
80. Kazlauskas RJ (2005) Curr Opin Chem Biol 9:195–201
81. Aharoni A, Gaidukov L, Khersonsky O, Mc QGS, Roodveldt C, Tawfik, DS (2005) Nat Genet 37:73
82. Branneby C, Carlqvist P, Magnusson A, Hult K, Brinck T, Berglund P (2003) J Am Chem Soc 125:874
83. Boller T, Meier C, Menzler S (2002) Org Proc Res Dev 6:509
84. Lalonde J, Margolin A (2002) Immobilization of enzymes In: Drauz K, Waldmann H (eds) Enzyme catalysis in organic synthesis vol 2. Wiley-VCH, Weinheim, p 163
85. Bornscheuer UT (2003) Angew Chem Int Ed Engl 42:3336
86. Reetz M, Zonta A, Simpelkamp J (1995) Angew Chem Int Ed Engl 34:373
87. Khalaf N, Govardhan CP, Lalonde JJ, Persichetti RA, Wang YF, Margolin AL (1996) J Am Chem Soc 118:5494
88. Zelinski T, Waldmann H (1997) Angew Chem Int Ed Engl 36:722
89. Lalonde JJ, Govardhan C, Khalaf N, Martinez AG, Visuri K, Margolin AL (1995) J Am Chem Soc 117:6845
90. Cao L, van Rantwijk F, Sheldon RA (2000) Org Lett 2:1361
91. Dyal A, Loos K, Noto M, Chang SW, Spagnoli C, Shafi KVPM, Ulman A, Cowman M, Gross RA (2003) J Am Chem Soc 125:1684
92. Cao L, Bornscheuer UT, Schmid RD (1999) J Mol Catal B 6:279
93. Dekker RFH (1989) Appl Biochem Biotechnol 22:289
94. Fernández-Lafuente G, Terreni M, Mateo C, Bastida A, Fernández-Lafuente R, Dalmases P, Huguet J, Guisan JM (2001) Enzyme Microb Technol 28:389
95. Terreni M, Pagani G, Ubiali D, Fernández-Lafuente R, Mateo C, Guisan JM (2001) Bioorg Med Chem Lett 11:2429
96. Rocchietti S, Urrutia ASV, Pregnolato M, Tagliani A, Guisan JM, Fernández-Lafuente R, Terreni M (2002) Enzyme Microb Technol 31:88
97. Sieber V, Martinez CA, Arnold FH (2001) Nat Biotechnol 19:456
98. Wong TS, Tee KL, Hauer B, Schwaneberg U (2004) Nucl Acids Res 32:e26

Author Index Volumes 51–100

Author Index Volumes 1–50 see Volume 50

Subject Index